Sonic Technologies

Sonic Technologies

Popular Music, Digital Culture and the Creative Process

ROBERT STRACHAN

Bloomsbury Academic
An imprint of Bloomsbury Publishing Inc

B L O O M S B U R Y
NEW YORK • LONDON • OXFORD • NEW DELHI • SYDNEY

Bloomsbury Academic
An imprint of Bloomsbury Publishing Inc

1385 Broadway
New York
NY 10018
USA

50 Bedford Square
London
WC1B 3DP
UK

www.bloomsbury.com

BLOOMSBURY and the Diana logo are trademarks of Bloomsbury Publishing Plc

First published 2017

Library of Congress Cataloging-in-Publication Data
Names: Strachan, Robert, author. Title: Sonic technologies: popular music, digital culture and the creative process / Robert Strachan. Description: New York, NY: Bloomsbury Academic, 2017. | Includes bibliographical references. Identifiers: LCCN 2016029249 (print) | LCCN 2016036146 (ebook) | ISBN 9781501310614 (hardback: alk. paper) | ISBN 9781501310621 (pbk. : alk. paper) | ISBN 9781501310638 (ePDF) | ISBN 9781501310645 (ePUB) Subjects: LCSH: Popular music–Philosophy and aesthetics. | Popular music–Production and direction. | Computer music–History and criticism. | Electronic music–History and criticism. | Digital audio editors. | Creation (Literary, artistic, etc.) | Music and technology. Classification: LCC ML3876 . S77 2017 (print) | LCC ML3876 (ebook) | DDC 786.7/16411–dc23
LC record available at https://lccn.loc.gov/2016029249

ISBN: HB: 978-1-5013-1061-4
PB: 978-1-5013-1062-1
ePDF: 978-1-5013-1063-8
ePub: 978-1-5013-1064-5

Cover design by Daniel Benneworth-Gray

Typeset by Deanta Global Publishing Services, Chennai, India
Printed and bound in the United States of America

For Marion and Kyle

CONTENTS

ACKNOWLEDGEMENTS

I would like to thank past and present colleagues at the Institute of Popular Music, University of Liverpool, and in the Department of Music at the University of Liverpool, notably Sara Cohen, Holly Rogers, John McGrath, Kate Smith, Simon Stafford, David Horn, Philip Tagg, Hae-Kyung Um, Richard Talbot-Leake and Nedim Hassan. I am especially indebted to Phil Davies for initially suggesting that I reflect upon my own creative practice, to Anahid Kassabian for numerous helpful conversations and suggestions, to Phil Kirby for stimulating conversations, his wealth of 'tech' knowledge and good humor, and to Mike Jones, Freya Jarman and Michael Spitzer for reading drafts of chapters and providing invaluable and insightful feedback. Thanks also to Phillipe Le Guen for extremely pertinent comments on an earlier article (published in French in the journal *Réseaux*) which informed Chapters 2 and 3 of this book and to Serge Lacasse and the anonymous readers who contributed greatly to later drafts of the manuscript. I would also like to thank my undergraduate students at the University of Liverpool, especially those who took my Music, Technology and Society, Sound Studies and Contemporary Genres modules over the past few years. The lively discussions and debates raised in these classes consistently made me reflect on the issues that became central to this book.

I am grateful to the team at Bloomsbury; Ally Jane Grossan, Michelle Chen and Leah Babb-Rosenfeld, for their support and patience during this project and to the many producers and artists who I spoke to during the research process including Sam Shackleton, Matthew Barnes, Sean Canty, Matthew Herbert, Donnacha Costello, Kamal Joory and Ruaridh Law.

Special gratitude goes to the Hive Collective: Matt Smith, David Sorfa, Alex Spiers, Bob Wass and Sam Wiehl for giving me the opportunity to be involved in such amazing creative projects, for introducing me to a massive variety of great music, for untold hours of useful conversation that underpins this book, and most of all, for being such a supportive, funny and understanding group of people. Without your friendship and collaboration this book would not exist.

Lastly, and most importantly, I am eternally indebted to my wife Marion Leonard, whose love, intelligence and understanding has been constant. This book, and everything else, is for you.

Introduction

In April 2009 the Apple corporation announced that it had sold 2.94 million Macintosh computers in the previous financial quarter, a 33 per cent year-on-year increase and a key factor in the best quarterly revenue and earnings in the company's history. A week earlier, Lady Gaga's 'Poker Face' hit the number one spot on the US Billboard Top 100 chart. The relentless exponential growth of the personal computer, even in the face of the most severe economic downturn since the 1930s, was being soundtracked by a hit recorded and produced in its entirety on an Apple MacBook laptop. This moment is emblematic of a tipping point in the relationship between digital technologies and creativity within popular music culture. Simultaneous developments in music and computer technologies, which since the 1980s had seen significant shifts in music production, were now ubiquitous and at the very heart of the recording industry. From established backroom star-makers creating multi-million selling tracks for the global market, through to amateur bedroom producers and artists operating within niche market or self-declared 'underground' scenes, the personal computer had increasingly become *the* locus for creativity.

The success of 'Poker Face' also occurred at a crucial point in terms of the consumption of popular music. The way in which the record was heard was indicative of how much listening patterns had changed in the previous decade. It would go on to be the most profitable digital track of a year when digital downloads accounted for a quarter of global music sales for the first time (IFPI 2010, 4) and was by far the most popular song on the newly launched streaming service Spotify.[1] After a decade of crisis the increasingly diversified major music companies were beginning to adapt to the new realities and reduced economies of scale afforded by the impact of the internet. The emergence of streaming further consolidated a movement towards an on-demand or cloud-based music economy that

[1]https://open.spotify.com/user/spotify/playlist/2B2cUq8sJVMHX7HsxC7zmR (accessed 28 February 2016).

continues to grow in significance,[2] and the mounting importance of single track downloads would have a major impact upon patterns of investment within the portfolios of large music companies.

What this example illustrates is how both in terms of production and consumption, the personal computer (and its unbundling into mobile hardware) has become the defining sonic technology of our age. From composition, recording and production to distribution, communication and promotion, digital technologies now play a central part in how we listen, how music is commoditized and what creative individuals do. This book examines these significant developments by focusing upon how digital recording and production technologies have had a transformative effect upon musical creativity. Taking in a broad range of digitally produced music, from globally successful pop through to electronic dance music and more experimental forms, it argues that recent developments in computer technologies and digital culture have been central in profound transformations in the creative practices, aesthetics and political economies of popular music.

Sonic Technologies is part of an emerging body of scholarship which has sought to engage with these developments. Recent book-length studies have provided insightful explorations of the effects of digitization on individual aspects of popular music. Brøvig-Hanssen and Danielsen's (2016) historical case study approach, for example, gives us a detailed textual understanding of how digital technologies have had profound effects upon the aesthetics of popular music since the 1990s. Rogers (2013) offers a convincing account of how the music industries have responded to digitization. Through its concentration on creativity this book seeks to offer a holistic approach which simultaneously takes into account the political economy, creative practices and textual conventions of contemporary popular music in relation to their convergence and mutual effect.

Digitization/digitalization

What follows is primarily an exploration of the effects of the digitization of popular music practice and the digitalization of the institutions central to its production and consumption. Digitization and digitalization are two closely related concepts that are so inherently bound up with each other that they are often used interchangeably or in a somewhat opaque manner. For example, within popular music studies scholarship the terms 'digitization' (Sandywell and Beer 2005; Jones 2012) and 'digitalization' (Fairchild 2008; Stahl and Meier 2012; Klein et al. 2016) remain undefined but are used to refer to a similar set of processes and contexts relating to the increasing

---

[2]In 2014 streaming revenues had taken over digital downloads in 37 international markets and now account for 32 per cent of global digital revenues (IFPI 2015, 7).

reliance upon digital technologies in patterns of production, distribution and promotion within the music industries. However, for the purposes of this book it is worth briefly unpacking the specificity of each.

In terms of its core application, digitization refers to the process of converting any type of information into digital form (see *OED* 2015b). This rather prosaic definition suggests a simple transfer from one medium to another. However, the transformation of data from analogue to digital form is never an innocent, purely archival transfer. For example, Ernst argues that even in its most perfunctory application, the archival storage of analogue materials, the process of conversion is inherently transformative. Rather 'than being a purely read-only memory, [through the process of digitization] new archives are successively generated according to current needs' (Ernst 2013, 81). In other words, the process of digitization in itself provides conditions for change, malleability and reorientation. Hesmondhalgh (2007, 243) points to the fact that the major components of cultural expression can be digitized (i.e. converted into binary code), making them more accessible, copyable and ripe for manipulation than before. In other words, different media are potentially interconnectable and changeable through their transformation. The idea of transfer and transformation is central here. A key argument of this book is that the transfer of tasks within the creative process from analogue to digital technologies fundamentally changes their nature, how they are perceived and carried out. Similarly, the actual transfer of sound recordings into digital files which are easy to upload, share and stream has had a fundamental effect upon listening practices, the economies of music and even the ontology of the musical work. With this in mind, within the context of this book, the digitization of music refers to the pragmatic act of transference from analogue to digital and the functional orientation of popular music practice within a range of digital technologies: principally, the integration of studio technologies within the personal computer and the centrality of the internet and Web 2.0 in the distribution and consumption of music.

In its *OED* definition, digitalization has a slightly different orientation, referring to the 'adoption or increase in use of digital or computer technology by an organization, industry, country' (*OED* 2015). This nuanced difference in meaning is worth hanging onto as digitalization implies a series of cultural and economic changes resultant from the collective adoption of digital technologies within a particular social group. As Brennen and Kreiss (2014) note, thinking about digitalization is less concerned 'with the specific process of converting analogue data streams into digital bits or the specific affordances of digital media than the ways that digital media structure, shape, and influence the contemporary world'. Digitalization can thus be thought of as encompassing the 'structuring of … diverse domains of social life around digital communication and media infrastructures' (ibid.). The application of the term 'digitalization' in the context of this book therefore, is concerned with the ways in which the institutions (businesses, scenes and networks) of music and creative individuals have increasingly changed and adapted their central practices in the wake of digitization.

The digital era and music technology

As an adjunct to this conceptual nuancing of terminology it is also necessary to undertake some historical clarification with regard to the timeframe in which digitization and digitalization have taken, and continue to take, place. As Théberge (2015) argues, rather than being regarded as a 'revolution', digitization should be viewed as a relatively long transformational process stretching back over half a century. Taking this into consideration I suggest that there have been three main epochs of digital music technology: that is, exploratory, expansive and convergent digitization. Théberge notes that the digitization of music can be traced back to collaborative experiments involving scientists and musicians in the foundational days of mainframe computing from the 1950s to the 1970s. It was at this stage that early algorithmic models for sound synthesis, along with computational and statistical methods of composition, were developed. These initial developments were both costly and tentative. As such, they took place 'outside of the immediate pressures of industry and commerce' in the context of large organizations such as Bell Laboratories in the United States, publicly funded research facilities such as the Institut de recherche et coordination acoustique/musique (IRCAM) in Paris and major research universities such as MIT and Stanford (Théberge 2015, 330). This era can be considered as the exploratory stage of the digitization of music; a period where ideas and technologies were coalescing in the somewhat esoteric and rarefied context of late Modernist art music and the 'research for research's sake' environment of academia. As such its impact tended to be limited to these distinct institutional environments, although they would provide the basis for the exponential growth and impact of digital technologies across other musical fields in the coming decades.

The second, or expansive, period of digitization took place from the early 1980s to the mid-1990s and was characterized by a rapid digitization of analogue instrumentation and studio technology and a resultant progression in common practices in composition, recording, sampling and sequencing across a broad-ranging context of music making and music production. This was largely facilitated by the financial and research investment of a number of large corporations such as Roland, Korg, Yamaha and Kawai. As Théberge (2015, 330) points out, the parallel release of Yamaha's DX7 (the first commercially available FM synthesis-based digital synthesizer) and the simultaneous introduction of the Musical Instrument Digital Interface (MIDI) protocol in 1983 constitute a watershed moment in the digitization of music.

MIDI provided a way to allow differing pieces of musical technology (such as synthesizers and drum machines) to communicate and synchronize with each other. The introduction of MIDI can be seen to have multiple effects in commercial, creative and aesthetic terms. First, there was the vast expansion of the music technology market, at both the professional and amateur level as differing types of musical equipment became interoperable (see Théberge 1997, 83–90). The introduction of MIDI also saw an increasing move from real-time recording towards the integration of pre-programmed MIDI-controlled parts as major components of many recordings through sequencing. This meant that there was a major change in the ways in which many recordings were made. Sequencing is a way of recording control data that can be read by electronic equipment to make it perform a given task (such as pitch, rhythm, volume, velocity, changing a waveform, etc.). Although sequencing technology existed before the advent of MIDI, the 1980s saw this equipment becoming easier to use, much more widely available (due to a substantial drop in price) and therefore increasingly utilized across many different types of recording. No longer were multitrack recordings solely the fusion of individual performances of musicians (albeit played at different times) mixed into a coherent whole. The widespread adoption of sequencing facilitated by MIDI meant that significant parts of recordings had never been 'played' in a traditional sense by any one individual. As Goodwin (1992, 263) reflected at the end of the 1980s, 'the most significant result of the recent innovations in pop production lies in the progressive removal of any immanent criteria for distinguishing between human and automated performance'.

This period of digitization also saw the widespread adoption of sampling techniques within the creative practices of popular music. Sampling has had a long musical history stretching back to the use of tape loops in *musique concrète* and minimalism. It was also at the heart of previous technologies such as the Mellotron, the computer music melodeon and the Fairlight CMI (see Harkins 2015). However, it was the digitization of the practice and the subsequent manufacture and marketing of discrete units by music technology companies that provided the catalyst for their widespread adoption. The Japanese manufacturers Akai were particularly important here, releasing a range of relatively affordable samplers from the mid-1980s onwards. The capabilities afforded by the digitization of the sampling process made sonic material much more reusable, malleable and open to transformation. As Katz (2004, 139) notes, digitization allowed for tempo and pitch to be increased or decreased in any increment, sounds to be 'reversed, cut, looped and layered … and certain frequencies within a sound can be boosted or deemphasized'. He argues that the transformative nature of digital sampling in itself served to 'transform the very art of composition' (2004, 157).

The widespread use of sequencing and sampling technology also had a major effect upon the way records sounded and their rhythmical and musical feel in the 1980s. In some ways the limitations of these technologies (in terms of rhythmic programming) gave a particular feel to music that used sequencers. This was often congruent with existing or emergent styles. For example, Toynbee (2000, 97) notes that the emergence of new technologies echoed the structures and conventions of post-disco dance musics in that 'step time programming matched the intense musematic repetition at a constant tempo that was called for in dance' and that 'the unit of the step, generally a bar or half a bar, tended to correspond to the key musematic units of bass line, drum beat, short riff' that were central to dance music's internal structures. While such an observation is indicative of how 'technology and music technique, content and meaning generally develop together, dialectically' (Middleton 1990, 90), there is little doubt that the viability of using such technologies by a broad base of musicians and the sonic and organizational specificities embedded within them led to a number of hybrid forms and new developments within popular music aesthetics.

The mid-1990s signal the start of a third period, which can be identified as convergent digitization: a period when a number of different strains stemming from earlier patterns of digitization converge upon the singular site of the personal computer. This is not to say that convergent digitization should be seen as outside a longer historical trajectory of technological and cultural change. As Chapter 2 outlines, the integration of various tasks within the site of the personal computer was the result of broader technological developments in personal computing from the 1980s. However, the increase in processing speeds of personal computers, in combination with the onset of web-facilitated communication technologies in the mid-1990s, signifies the start of a period in which integrated computer production became increasingly more viable and access to production technologies by a large body of users was able to reach unprecedented levels. In terms of music technology this third period of digitization is characterized by a series of convergences: an increasing transfer of tasks from hardware to software, the integration of creative work within the singular site of the personal computer and the mounting importance of computer mediated communication within the cultural and economic contexts of popular music. These changes have had multiple and significant effects: the situation of production and creative practice within virtual environments, an increasing interoperability enabling production to transcend geographical location, a significant increase in the numbers of people involved in music production made possible by increasing access to studio technologies, and a lowering of the entry barriers to studio technology through cost reduction and ease of use.

It is these developments that form the core subject matter of this book. It suggests that the growth in the accessibility of personal computers has not only meant that production technologies are now available to a much wider caucus of individuals but that the gap between professional and amateur (in terms of equipment, knowledge, practice and sound) has significantly narrowed. It also examines how the incorporation of a number of hitherto separate tasks within the creative process into an integrated computer environment has resulted in a new set of creative modalities and a re-conceptualization of musical thought. These new modalities have also in themselves led to a transition in what it means to be creative. The lines between composition, production and performance have become progressively more blurred, leading to a whole generation of practitioners whose roles are less easily placed within stratified divisions of labour that traditionally characterized the creative processes of popular music throughout the twentieth century.

The book also argues that computer-based music production has increasingly brought the physical and textural properties of sound to the fore within the creative process. This foregrounding of the sonic is an ultimate fruition of trends within the way in which popular music has been made since the 1950s (see Gracyk 1996; Zak 2001) and was amplified by the expansive period of digitization in the 1980s. In the light of this, the book identifies a materialist approach to sound and the emergence of a multitude of differing articulations of the digital aesthetic across a wide range of popular music genres. Computer-based music production enables and demands that the user work directly with captured and generated sounds that are at a remove from processes and competencies of performance traditionally associated with musicianship. This is not to say that traditional modes and skills of performance are *always* sidestepped within the within the modalities of computer-based music production. Instead, the argument will be presented that beyond the capturing of sound, the computer environment needs to be interrogated for the way that it allows, encourages and facilitates the making, processing and manipulation of sound. In other words, the computer environment should not be understood as a neutral way of recording, capturing and presenting sound but as highly influential to the creative process in its design, construction and capability which in turn have a central influence on the sounds and eventual recordings that are produced.

Perhaps the most important technological development of this period of convergent digitization, and one that forms a key area of study within this book, is the coming of age of the computer-based digital audio workstations (DAWs). In contrast to the types of institution that drove development in the exploratory and expansive periods of digitization (publicly funded/educational and large technology manufacturers respectively), change in this era has tended to be facilitated through a combination of large computer hardware manufacturers, smaller more specialized software developers and the structural economic connections between them. Computer-based DAW software produced by companies established in the 1980s (Steinberg's Cubase, Avid Technology's Pro Tools) and tech start-ups of the 1990s and 2000s (Image-Line's Fruity Loops, Ableton's Live, C-Lab's Logic and various applications developed by Propellerheads, including Reason) have been crucial in the changing landscape of music production during the past two decades.

DAWs are all-in-one applications installed on computers, which provide a visual interface and collection of functions whereby recording, sound generation, editing and mixing are able to be undertaken within a singular virtual environment. In other words, all the work that goes towards the creation of new musical texts can be generated or controlled through the DAW. External instruments and microphones can be connected to the DAW (usually via a separate audio interface or sound card) whereby audio can be recorded digitally to the computer's hard drive. Sounds can be generated through the use of built-in or third-party 'virtual instruments' such as synthesizers and drum machines. Musical structures can be sequenced and edited through the arrangement of MIDI information and final mixes (whereby the levels of instrumentation are appropriately balanced) can be completed. The integration of these functions into a single interface (especially since the vastly increased processing speeds of personal computers in the 1990s allowed for smooth simultaneous handling of audio and other tasks) has served to fundamentally alter many types of musical practice and has also provided a singular but multipurpose tool that has become dominant in the production contexts of a significant proportion of popular music creativity.

Actor network theory and affordance

Yet, digital technologies are more than mere tools; they are active and enmeshed within the creative process in a central way. In actor network theory terms, they are a key part within the socio-technological-human networks that go towards making music in the post-digitization environment. Actor network theory (ANT) views both humans and objects as non-hierarchical actors (or actants) within sociotechnical networks. Both have agency, and both are negotiations of the social and the technical. Latour's (2005) formulation of human-technology interrelations suggests a diverse network of human and non-human agents (people, machinery, technologies and objects) that combine to make achievable what neither could accomplish individually. For Latour, taking all these elements into account avoids a reductionist tendency within 'sociologism and technologism' by acknowledging that we are never just faced with 'objects or social relations, we are faced with chains which are associations of humans ... and non-humans' adding that 'no one has ever seen a social [or technical] relation by itself' (Latour 1991, 110). ANT provides a useful framework in that it 'alerts us to how the technical and the social are inextricably linked, in turn sensitizing us to the fact that instruments and associated devices are not passive intermediaries but active mediators' (Prior 2008b, 315). In other words, there is a dispersal of agency across differing biological and non-biological sites. In turn, the particular make-up of a network serves an essential defining role in what it ultimately produces. Furthermore, each network of creativity in this regard (for the purposes of this discussion a particular type of technology, a human producer and the particular field in which they work) is distinct and relational according to the particular make-up of a given network. Seeing technological/human relationships in this way accounts for the 'manner in which relationships in the real world multiply, overlap, and change [calling] attention to the motile web of relations that define and enable any actor's role. The network affords an actor certain ways to work; change the network, and you change the actor' (Piekut 2014, 194). This is essential to understanding the approach to creativity within this book. While this book is not an ANT study, one of its fundamental arguments is that the virtual domain of the digital production environment intersects with the creative individual and their position in relationship to the economic and cultural context in which creativity unfolds. Within this network technology affords certain uses and trajectories that are always mediated by other points in the network.

In order to capture this complexity, one of the key theoretical threads of the book is an ecological approach to creativity. Informed by the work of James Gibson (1966, 1978) ecological psychology has had a wide-ranging influence across studies of perception within a number of fields relevant to this book, including human–computer interaction (HCI) (Norman 1988, 1999; Bertelsen 2006; Turner 2008), music (Winsor 2000; Oliveira and Oliveira 2002; Clarke 2005; DeNora 2005; Krueger 2010), music education (Pea 1993; Gall and Breeze 2005; Reynolds 2008) and, indeed, ANT itself (Latour 2005, 72). However, although there is unambiguous benefit from the theoretical insights of the ecological approach to understanding the intricacies of musical creativity, work that takes this approach is only beginning to emerge (Strachan 2012, 2013), including Zargorski-Thomas's (2014) pertinent application of ANT and affordance theory to record production. A theoretical constant across this work has been through the concept of affordance, a term coined by Gibson (1966) to describe how perception operates within a relational structure between organism and object. Within this theorization there are certain actionable properties that are latent within an object or environment which may be acted upon by a human or animal. Objects thus *afford* a range of uses (various but not unlimited) that are perceived by a human actor but are subject to particular subjectivities or socialization. This book uses this core concept in two ways: first, to examine how the interfaces of digital technology are based around conventions that offer certain affordances which frame creativity and second, how particular sounds offer certain uses (sonic affordances) which are at the heart of the musical choices that are made during creative work.

Technology, creative practice and experience

Implicit within the idea of creativity as a network is that technologies are active in the production of experience for the human actor. What this book also suggests is that any understanding of creativity within contemporary popular music must take into account the experiential relationship between the individual and technology. Experience lies at the heart of how this book approaches technology and the creative process. Again, we should understand technology as being much more than simply a tool here. For the creative individual, interactions with technology provide experiences that have affective, emotional power and are entwined with cerebral processes and bodily sensation. In this sense the book is informed by my own personal experience and reflections. As an academic who also makes music and sound art with digital technologies, the idea of experience was central in guiding my thinking towards our relationships to music production technology. I have worked on a series of musical and sound art projects since the early 2000s and, as somebody who also teaches and researches in the area of popular music, I perhaps quite naturally began to reflect upon my own praxis. Yet, I found there was something of a disjuncture. The major part of my life was given over to utilizing social and cultural theory (informed by sociology, anthropology, popular music studies and cultural studies) in order to examine the processes, practices and texts of popular music. At the same time, my creative life as I experienced it seemed rather disconnected from these deconstructive and analytic tendencies. As I worked in my studio, collaborated with creative colleagues and performed in club events and gallery spaces, I experienced moments of pleasure, immersion and transcendence that seemed marked off from everyday life, instinctive and highly *personal*. To an extent, this book is an attempt to meld and reconcile these two areas through a line of enquiry that seeks to explore the social and technological construction of the creative experience. With this in mind, throughout the book I have used field notes from my own experiences of doing creative work, and experiencing the creative work of others, as phenomenological jumping-off points for the discussion of various key themes and subject areas. By doing this, I am not seeking to suggest that general patterns can be extrapolated from specific and often very personal individual moments from within the creative process. Rather, the converse is the case. These individual recollections of experience are presented here as illustrations of how general schemas relating to digital technology and music become embedded within personal experiences of the creative process and the reception of music.

Given the concentration on the experience of the creative process within this book I have also attempted to account for the voices of creative individuals throughout. These accounts are drawn from a variety of published sources and from my own interactions and interviews with electronic music producers. While again, these voices are not presented as objective, exemplary proof of more general trends, they do provide an empirical set of examples of how individuals experience and understand their own actions within the social-human-technological networks of creativity in nuanced and personalized ways. In a more general sense they are included to help situate an understanding of music technology that is firmly rooted in the experiences of it users. As Taylor (2001, 34) points out, 'everyday people and how they use everyday objects of technology ... to make, disseminate and listen to music' have generally been missed out of science and technology studies (particularly in ANT) in favour of 'agents of change', the 'inventors, innovators and engineers'. As such, the voices of creative individuals are an attempt to come to terms with what Mcarthy and Wright (2006) term the 'felt-life' of technology (the way in which technology provides experiences of engagement, flow, irritation and fulfilment) in a way that has perhaps been hitherto absent.

Historical context

Despite this thread relating to the personal that permeates sections of the book, the social is always present. The historical context in which changes in creativity take place should never be overlooked. Our relationships to technology change and unfold over time and as such must be situated according to their particular place in history. Sections of this book (particularly Chapters 2 and 5) are broadly historical in nature, seeking to address the specificities of change and examine the processes that lead to change. While these chapters examine individual historical trajectories (the development of the DAW and the emergence of digital aesthetics since the 1990s respectively) the broader historical argument that underpins the book is twofold. First, the book argues that the period of convergent digitization since the 1990s engenders particular types of technological/social/human assemblages that guide and shape the creative experience in distinct ways relating to this specific historical context. Secondly, I want to suggest that an examination of digital production practices and technologies is key in understanding the cultural and economic transformations that have happened in the wake of digitization.

The changes in creative practice outlined in this book take place in the context of, and are intimately entwined with, seismic changes within music distribution, consumption and economy. The way that consumers access music has gone through a period of rapid change in the past two decades. The move from physical products to downloading and streaming happened in such a relatively short space of time that the period from the turn of the millennium has felt like an era of continual flux and uncertainty for the music industries. What is certain is that the sharp falls in demand for physical product, which had been the main revenue generator for the recording industry in the twentieth century, have meant that large music companies have had to fundamentally adapt their practice in order to maintain the power relationships that have traditionally maintained their dominance within the marketplace. To give a sense of this change, in 1999 the trade value of physical recorded music had reached $27.3 billion; by

2014 it was worth just a quarter of that value ($6.8 billion) (Klein et al. 2016, 4). There have been several responses to this significant decrease in value. One aspect has been the increasing importance of the live music sector and music publishing through the exploitation of synchronization rights as a potential replacement source of income for writers and record companies (Williamson and Cloonan 2013) and concurrent strategies such as the 360° contract, designed to derive income from all aspects of an artist's income (Stahl and Meier 2012; Jones 2012). Another has been through the existing music companies diversifying into aggregators such as streaming and download services through a series of joint investments and the negotiation of preferential business deals (see Chapter 1).

A related and key effect of these developments has been a transformation in the way sound recordings are monetized. The decline of physical album sales has seen a necessary concentration on individual tracks. As Klein et al. (2016, 4) note, the 'displacement of brick-and-mortar record shops by iTunes and other digital retailers hastened the "unbundling" of the album and the return of the single, a shift that fed the fragmentation of recording revenues'. Marshall (2013, 65) points out that since digital sales were included in IFPI statistics, singles sales actually showed a significant increase in the following five years up from $346 million in 2004 to $1.636 billion in 2009. As Marshall concludes, 'The problem for the labels is that people are choosing to consume music in smaller chunks, and thus most of the transactions are actually quite small and less profitable than CD purchases' (ibid.). In the period since these statistics, streaming has become the emergent revenue stream rising from just 7 per cent of global digital revenues in 2010 to 23 per cent in 2014 (IFPI 2015, 9) while single downloads stabilized somewhat (1.904 billion US dollars in 2014, ibid.). Given that a majority of what is streamed is a tiny proportion of the tracks available (*The Economist* (2009), for instance, estimates that the most popular 5 per cent of tracks on Spotify account for 80 per cent of all streams), both of these trends point to the same thing: a concentration on individual tracks within the recording industry and the heavy investment in a small number of globally marketable stars.

In turn, these knock-on effects of digitization have resulted in a number of dominant stylistic trends which are highly relevant to this book. As I write (September 2016), the week's top 100 most played Spotify tracks are almost entirely dominated by electronic music production. If this is taken as a yardstick for contemporary mainstream pop music, then it is a soundscape dominated by drum machines, 1990s-inflected synthesizer pads and bass tones, highly stylized vocal staging and genre indicators of dance music such as filter sweeps, pumping side chain compression and breakdowns. Even most traditionally rock-based acts that were in this selection such as Ed Sheeran and Coldplay utilize electronic production styles in order to orientate themselves towards the mainstream market.

This snapshot of contemporary popularity is indicative of a new mainstream in which electronic music and electronic production techniques (and, in particular, DAWs) have become progressively more central within the global popular music market. Toynbee (2002, 122–3) sees mainstream music as a rather self-evident generic grouping and 'mainstreaming' as a social process in which the music industries 'utilize musical texts and generic discourse which "fold difference" in and articulate distinct social groups together' towards an 'aesthetic of the centre, a stylistic middle ground'. Recording technologies have always been at the heart of this mainstreaming tendency. Positioning an artist or particular type of music towards mass audiences has always been a process of using production style to make recordings fit with common institutional expectations (such as radio formatting) and contemporary audience expectations about production. Since the mid-2000s this has taken the form of a hybridization of a number of genres in the context of the digital production environment. The creative intermediaries that are central in the making of mainstream pop have increasingly incorporated the formal and timbral characteristics of electronic dance musics into their work along with a parallel cross-fertilization of hip-hop, R&B and traditionally white pop idioms. The most commercially successful end of the current market is therefore now dominated by a relatively small number of hybrid songwriting/production teams who have utilized the flexibility and interoperability of the DAW within their creative processes. The DAW is now the site of songwriting, tracking and production for a significant section of cultural producers within the upper echelons of the contemporary pop industry in a move away from the traditional creative context of the large commercial recording studio (see Chapter 1).

The issues raised in this book, then, are highly pertinent to practices which have become centrally embedded within the recording industry in its move towards digitalization. Both digital production techniques and electronic music are now at the very heart of popular music's mainstream. This is illustrated by another significant development of the past decade which has seen electronic dance music (EDM) itself becoming an ever more significant presence within the global popular music market. A combination of the catch-all rebranding within the US market of dance styles popular in Europe for decades as EDM (Jopson 2015), the conscious matching of established dance music DJs with mainstream pop stars for one-off collaborations in order to crossover to a larger sector of the market (Mortensen 2012) and major investment in live dance music events have seen a major repositioning of EDM across the music industries and an exponential growth in its financial worth. In 2014 the global EDM business had a value of US $6.9bn, up 12 per cent from $6.2bn in 2013 (Watson 2014, 84). In the United States, Soundscan data indicated that sales of recorded music for EDM grew 8 per cent in 2013, a year when every other genre posted declines, including a 12 per cent decline in rock sales (ibid.).

With the significant decrease in album sales and the concentration upon individual tracks by the recording industry, the emergence of EDM as a major source of income makes logical sense. During the first emergent era of dance music in the 1970s and 1980s, genres such as disco, house and techno were seen as areas for short-term investment, licensing and ultimately, secondary to core rock and pop artists who were allowed a more sustained investment and development program within the major recording companies. The fragmentation of the market and move away from albums mean that EDM artists and producers are now seen as much more central within the portfolio of the major music companies. EDM is track based, globally marketable, and has significant licensing and third-party brand tie-in potential across international territories with little in the way of investment in terms of recording costs. In addition, the significant investment into EDM festivals and events by large companies such as SFX and Live Nation has substantially increased both the market for the genre and the visibility of its DJs and artists (Watson 2014, 18). Global stars such as Calvin Harris, David Guetta, Skrillex, Tiesto and Deadmau5 now have enormous market recognition and earning potential. For music companies EDM allows them to gain access to the increasingly significant financial rewards of live events through 360° deals and also to piggyback upon the existing market capital of established DJs and dance acts. As Millard (2013) argues, EDM 'seems clinically designed for a post-album sales era, a rebranding … that communicates with it an entire new template for commerce, the single providing a much lower barrier to entry to getting to the actual music, serving as a much more baldly obvious commercial for the live show, which is the overt crux of the EDM experience'.

These developments in the mainstream reveal fundamental issues about both common creative practice and a broader strain in strategic thinking within large music companies. In his work on the multinational recording industry of the 1990s, Negus points to the organization of investments of financial, promotional and managerial resources into popular music artists based around genre. His findings showed that through portfolio management certain types of artist were privileged as priority acts suitable for long-term investment and support (white rock acts, globally marketable pop stars) while others were regarded as cash cows ('which can produce sizable profits … with [only] minor modifications and modest ongoing investment' 1997, 48) or flash-in-the-pan trends to be capitalized upon quickly within a limited timeframe. To a large extent the wake of digitization has seen the large music companies becoming more reliant on cash cows and flexible and relatively immediate responses to trends in the market, as opposed to core long-term, and comparatively expensive, investments. Currently, widespread practices such as placing emergent artists with established stars through guest appearances and matching priority artists with songwriting/production teams that have a current track record of hits for individual tracks allow large music companies to flexibly capitalize on commercial currency across their portfolios in an immediate assessment of the contemporary market. As Chapter 1 illustrates, these trends are in themselves facilitated by the flexibility, interoperability and geographical transcendence afforded by digital production technologies.

Outside of these developments in the mainstream a contemporaneous countercurrent of this recent historical context also forms a central point of study within this book: the proliferation of niche market genres that find small but global audiences through digital technologies. As Bennett et al. (2006, 2) note, the period from the 1990s has seen an increasing 'fragmentation of market, styles and consistencies' facilitated by processes of globalization in tandem with 'significant technological shifts in production and consumption'. The book also considers issues relating to creativity and stylistic change from within this milieu. It draws on examples from self-declared 'underground' and experimental electronic genres such as electronica, dubstep, techno, hauntology and vaporwave to examine how the effects of digitization have been pervasive across all levels of contemporary production and creativity. This simultaneous assessment of both the macro and micro contexts of popular music production is illustrative of the profound and far-reaching implications of the issues raised in the following chapters. In short, the economic and cultural significance of electronic music and electronic production techniques in the current climate means that the subject matter and issues raised in this book are to a large extent at the heart of contemporary popular music culture.

Organization

To summarize, *Sonic Technologies* offers a critical analysis of important contemporary developments in digital culture and examines how digital technologies have impacted upon music production and musical thought. It develops an understanding of how musicians think about and engage in creative practice within contemporary popular music culture by offering an ecological approach to creativity and situates broader contemporary cultural production within recent developments in technology and culture. It examines how advances in musical technology, the spread of ubiquitous personal computing and the democratization of creative digital music applications have meant that for a significant number of musicians, working directly with sound has become the main focus for musical creativity. This book seeks to explore how creativity works as an experience within this new digital context.

The book also examines a series of structural and social issues. It questions how far the proliferation of computer-based digital production technologies and the ubiquity of the internet might be read as a significant democratization of musical knowledge and music technology. It examines how the contexts of popular music creativity have transformed in structural and economic terms. It considers how factors such as an assessment of social use, listening environment and socially constructed notions of creativity affect the ways musicians and producers use technology. Lastly, it examines how the integration of digital technologies within our daily lives has had a profound effect upon the aesthetics of popular music.

The book is organized thematically through five chapters. Chapter 1 provides a broad contextual overview from the 1990s to the 2010s relating to the effects of digitization on the production and distribution of popular music. It first describes how the dual developments of virtual studio technology within the computer environment and the emergence of the internet have led to an opening up of production practices to a much wider base of users. It argues that the availability of free and low-priced music production software and increased access to information have allowed a lowering of the entry barriers to music production. The chapter goes on to examine how, despite this perceived democratization of technology, large music companies have reacted to digitization by seeking to maintain their control over the market by adapting their structural organization and investments accordingly. The chapter concludes by describing how digital technologies have been incorporated increasingly centrally into the production practices of mainstream pop music through the development of new and dominant patterns of creative labour.

Chapter 2 provides a historical narrative that traces the entwined development of the digital audio workstation and the emergence of the personal computer as a ubiquitous consumer technology from the 1980s. It argues that the development of overarching design conventions within personal computing has had a profound effect upon how music technologies became integrated within the site of the personal computer. It traces how from the earliest development DAWs were designed in relationship to an incipient logic of 'intuitiveness' and interoperability that had a fundamental effect upon how sound is ordered and organized within the DAW environment.

Chapter 3 builds upon this narrative by providing an analysis of how creativity is structured within the virtual environment according to constraints and possibilities afforded by certain design logics. It argues that technology offers certain structures that are essential in understanding commonalities within the creative processes of electronic musicians, songwriters and producers.

This line of enquiry is expanded in Chapter 4 which attempts to place creativity in a broader sociocultural framework. First, it examines how the creative process is driven by certain socially constructed discourses of creativity before describing how musicians and producers engage with the social uses and institutional frameworks of popular music.

While equally concerned with the effects of computer–human relationships in the digital age, the final chapter takes a slightly different tack. It examines how the integration of DAWs into the creative process has had material effect upon the aesthetics of popular music. It first outlines the emergence of a number of cyber genres, the emergence of which can be seen in the light of both production technologies and as an artistic response to the effect of digital technologies upon identity and subjectivity. The second half of the chapter uses the case study of a particular manifestation of virtual studio technology (Auto-Tune) which has become all but ubiquitous in the vocal staging of mainstream popular music.

CHAPTER ONE

Digital technologies, democratization and cultural production

Making music is getting more and more accessible. As Steve Jobs pointed out in his latest keynote speech, anyone can now be a music maker. Soundcloud's part is that, as well as being a really useful tool for the major artists and record labels, we empower this new generation of music-makers to share their creations and get feedback.

DAVE HYNES, VICE PRESIDENT OF SOUNDCLOUD.
(QUOTED IN SWASH 2011, 7)

This pronouncement from a senior executive of one of the most significant internet start-ups of recent years is typical of the way in which new media companies commonly frame their respective business models and USPs. For the new media industry, from high-tech corporations to start-up companies, the idea that their products somehow empower consumers, enable creativity and allow a voice for individual expression has become an overriding logic of product development, as well as a key to how such products are marketed to, and understood by, the public. Such a logic is a direct reflection of the level of ubiquity and extraordinary financial success of web-based services and apps that have become integrated in the lives of consumers from the mid-2000s onwards. From Facebook and Twitter to Instagram, Soundcloud, Vimeo and YouTube, these services are essentially platforms for users to create, upload and share differing types of content. These services have become *the* defining business models of Web 2.0 (generally accepted as the second stage in of internet development), a historical period characterized by the emergence of social media and a move away from static HTML pages towards more dynamic or user-generated content, further accelerated by the further ubiquity of the 'untethered' web – as experienced and accessed through tablet computers and smartphones. As such, the idea of the empowered user is entwined within the DNA of Web 2.0, defining its core functions and discourses, in which leveraging the power of the consumer has become central to the business models of tech companies. Simultaneously, the prosumer (a portmanteau of producer and consumer first coined by Alvin Toffler in his 1970 work, *Future Shock*) has become recognized as a target market for technology, computing and information technology industries and is also a contemporary cultural buzzword (Gerhardt 2008; Gunelius 2010). Prosumer technologies such as camcorders, sound recording equipment and DAWs have provided relatively affordable tools which have been utilized in an uncountable plethora of high-quality, self-produced media.

These twin aspects of digitalization have meant that the democratization of culture, empowerment and everyday creativity have become enduring concepts in common public discourse about internet culture and digital technologies more generally. In the decade since *Time* magazine (Grossman 2006) famously pronounced 'You' (as a signifier for the general public) as their person of the year for 'founding and framing the new digital democracy' such discourses have become so pervasive that they have become almost self-evident.

This broader context of technological progression and discursive framing is crucial in coming to terms with the shifts in music production practices facilitated by personal computer – based applications that are central to this book. And in some ways we can take the general thrust of these broader everyday discourses of democratization as highly apposite. There is no doubt that technological developments from the 1990s onwards have led to a levelling between the 'professional' and 'amateur' studio both in terms of achievable audio quality and the actual techniques and interfaces used to realize finished recordings. As Théberge (2012, 83) notes, hardware and software developers have increasingly designed professional and consumer versions of their products with almost interchangeable features, meaning that 'the sheer power of these technologies has largely fulfilled the dream of a professional quality home recording that earlier technologies ... only promised'. As a result 'the distinction between what can be considered a "professional" or "commercial" project studio and simply a "personal" or "home" studio ... [has] become increasingly difficult to make' (ibid.).

However, while these technological changes are undeniable and highly significant, I do not want to categorize them as unproblematically providing some kind of across-the-board democratization of musical practice. Rather, this chapter will argue that the effects and implications of these technological changes are complex, multifarious and subject to a variety of nuanced social and cultural contexts. In effect, they have to be situated within distinct and often separate cultural and industrial realms. As the wealth of sociological work from the culture of production perspective (Negus 1992; Peterson 1997; Du Gay 1997; Hesmondhalgh 2007) to Bourdieian analyses of the cultural industries (Pratt 1997; Hibbett 2005; Strachan 2007; Moore 2007; Scott 2012) has demonstrated, cultural production and creativity takes place according to specific discursive, economic and cultural conventions. We should not be blinded to implications of these existing and evolving frameworks by merely replicating the pervading utopianism of public discourse. With this in mind, I want to suggest that the concepts of democratization and the prosumer when applied to musical practice are somewhat knotty and less than straightforward than pervading popular discourse would suggest. Rather than taking as wholesale the idea that developments in technology have constituted a disruptive break with previous practice, it is important to place them within the structural specificities and historical legacy of popular music production.

In order to unpack this, the chapter will initially outline issues around digital democratization in terms of production and distribution. In particular, it identifies three often overlapping relevant fields: amateur/hobbyist, small-scale cultural production and the mainstream recording industry, and goes on to examine how digital recording practices operate within them. First, it addresses the question of access to technology by examining how the distribution of cracked software (whereby the removal of copy protection features from software makes it freely available) and a depreciation of the price of VST applications have broadened the number of people utilizing studio technology. It argues that this factor alongside the easy dissemination of information about studio skills and more intuitive user-friendly modality to VST technologies has served to lower the entry barriers to music production. The chapter then unpacks common discourses relating to the effects of digitization upon distribution and mediation, arguing that the wider dispersal of studio technologies does not necessarily equate a singular democratization of popular music practice. Rather, it suggests that digitization should be understood as having differential effects relating to the specificities of a given cultural field.

Secondly, it will examine how such technologies have significant implications for small-scale cultural production operating outside of, or on the periphery of, the mainstream. Finally, it will examine the situation and use of digital audio workstations in the mainstream pop industry. Through a detailed analysis of hit single recordings in 2015 it will argue that digital technologies have led to both changes in the sites of production of popular music and in the predominant creative units through which pop is produced.

The democratization of production

In a series of articles on the British post-punk, indie and dance music scenes Hesmondhalgh (1997, 1998, 1999) draws upon various theoretical positions to propose a typology of democratization as a way of assessing 'the level of democracy in any given form of media production' (1998, 256). These elements of democratization are: increased participation and access, a decentralization in terms of media organizations and technologies, equality in levels of reward and status for participants and the emergence of innovative and diverse forms of expression. This typology provides a useful framework for a discussion of the effects of digitization upon cultural production. Although Hesmondhalgh's research discusses a period when the full impact of the internet on music cultures, production, consumption and economic flows was yet to be felt, each of these four elements has had material effects on the ways in which music practices and their promotion have developed over the past two decades.

The key thrusts of the democratization argument in relation to digitization and music practice have been twofold: the democratization of production and the democratization of distribution. So, on the one hand, we have the availability of digital music making technologies to a much wider group of consumers and producers (Jacobson 2011–12; Leyshon 2014) and, on the other, the possibilities of distribution of recordings and access to a wider audience outside of traditional media and industry channels (Adegoke 2006; Mali 2008; Swash 2011; Hracs 2012). In tandem these two factors are often read as producing a quasi-utopian virtual landscape in which the traditional barriers for entry to the music industry and the control of distribution have been eroded. Ryan and Hughes sum this position up neatly:

> Whereas technology previously alienated the average person from the music production process, the relationship is now reversed, and technology has returned the means of production to the people, ushering in an era of recording democracy. If you want a voice you can have one. And the audience votes with the click of a mouse. (Ryan and Hughes 2006, 240)

With regard to the democratization of production there is no doubt that recording, sequencing and synthesizer technologies have in the past twenty years become much more accessible to a wider range of people purely through the development of computer-based technologies and their widespread availability. First, software and hardware targeted at the consumer market (as opposed to the professional audio industry) has become much more sophisticated and have undergone a price drop in real terms. Secondly, the gap between professional- and amateur-targeted technologies has significantly shrunk in terms of the quality of recording that can be made on either. Thirdly, internet technologies have meant that a significant proportion of music software programs have been available as illegal downloads, thereby massively increasing the number and demography of consumers who have access to such programs.

This third issue is highly significant in that anybody with access to an internet connection and a personal computer, and having some level of computer literacy can engage in production at some level. This presents a substantial change from the hobbyist market outlined by Théberge (1997) in his study of home studio technologies of the 1980s and 1990s. The rapid permeation of MIDI technologies in the early 1980s led to a concurrent expansion of the consumer market for synthesizers, sequencers and home recording equipment. The entrance into the market of large Japanese corporations in the mid-1970s along with the advent of digital technologies and MIDI protocol led to a significant drop in price and a level of technological compatibility that enabled home studios to become widespread for the amateur and semi-professional musician (ibid.). However, to set up a home studio as a hobby nonetheless required a fairly significant financial outlay on hardware and appropriate space in which to house such studio facilities.

A new set of possibilities was opened up in the 1990s when the differing tasks of the home studio could be performed on the singular site of the personal computer. As Chapter 2 will discuss in detail, personal computers became more powerful and widely available in the 1990s and the types of sampling, sequencing and digital recording technologies that emerged in the 1980s became available to home recordists through newly developed computer programs in the form of virtual studio technology. The shift from hardware to software meant that the DAW provided a singular environment that allowed users to record, edit and play back digital audio on their personal computers in a process of convergent digitization. Programs such as Cubase, Pro Tools and Logic developed over a number of years to provide professional standard production facilities within virtual environments without the need for external hardware such as synthesizers, sequencers or tape.

Cracked software, peer-to-peer networks and warez

Increases in the processing speeds of computers were also directly paralleled by the emergence of internet technologies, and with these developments distinct cultures and discourses emerged. The internet provided a means for the dispersal and sharing of software and an overarching logic for such sharing to take place on a mass scale. The originating ethos of the internet as a channel for the free distribution of information has had a fundamental legacy in its legal and cultural history. In many ways, these ideas were built into the DNA of internet culture. As Turner (2006) convincingly argues, the countercultural roots and connections of central figures involved in the incipient technological hub of Silicon Valley served to shift the discourses of computer culture from its roots in post-war military research towards a collaborative technologically driven utopia deeply influenced by the communal ideals of the hippies. As a direct result of its emergence from within this milieu, a common discursive position within internet culture throughout its existence has taken the form of a popular 'intellectual property critique' (Hemmungs Wirtén 2006) as manifest in the pro-piracy/anti-intellectual property movements. Forums and file-sharing sites have become 'spaces where intellectual property and file-sharing are debated, wherein anti-copyright/pro-piracy rhetoric constitutes a kind of "party line" for this emergent audience' (Lobato 2011, 114). As such, there has been something of a normalization of file-sharing among internet consumers.

An element of this culture has been dedicated network of individuals who used hacking as a form of political activism (or on a spectrum from anarchists to neo-libertarians) through a commitment to removing copy protection features from computer software and making free copies widely available. From the late 1990s what became known as the warez scene became (in terms of software) the most important source of copyrighted material illegally distributed online (Décary-Hétu 2014). Based around a 'modern gift economy' with 'a consistent and internally rational structure of actively anti-economic behavior' (Rehn 2004, 359) the warez scene emerged as a network of individuals whose shared values and subcultural practice led to a rapid acceleration in the global distribution of cracked software. A key motivating factor for individuals involved in the scene is the accumulation of subcultural capital leading to an exponential growth in the amount of applications available for download. The unlocking of software through the circulation of stolen passwords, the generation of key generation software (which generate personal unique passwords) and the disabling of other security functions became a central form of subcultural activity within this virtual scene. The fact that being the first to crack newly released software gave hackers kudos, status and a relative amount of fame within the scene led to a very quick turnover of software from official release to pirated version (Rehn 2004). In addition, the distribution of cracked copies (usually bearing the hacker's digital signature) outside of the scene via popular torrent aggregators and file-sharing sites in order to increase the hacker's reputation and fame meant that illegal software became available to a much wider audience. As a significant part of this movement, the hacking and distribution of cracked music production software became a focus for a buoyant audiowarez subgenre within the warez scene, with hundreds of newsgroups, aggregator and torrent sites emerging in the period. This meant that virtually all major audio software packages became available for free within a few clicks of a mouse to anyone who had the inclination to look for them. As such, cracked software and peer-to-peer networks have been an important point of entry to production technology for a significant caucus of consumers. For example, Whelan's (2006) ethnographic study of amateur hip-hop and dance production points to widespread use of cracked copies by his respondents, indicating that for a large number of young musicians peer-to-peer networks constituted the 'basis of community' in which the use of illegal software had become an accepted part.

The impact of illegal downloads upon the music production software sector thus constitutes an unprecedented lowering of entry into studio technology. Although exact figures are difficult to come by, given the illegal nature of downloading, there are definite signs that cracked music software actually constitutes *the* major way in which computer-based technology is accessed by consumers. For example, the growth in computer hardware peripherals such as soundcards and control surfaces has substantially outstripped the equivalent growth in music software sales during the same period. Figures up to 2014 show a 35 per cent increase in the value of the soundcard market over a ten-year period while recording and sequencing software had experienced an 11 per cent decrease solely in the year leading up to the study (Challacombe and Block 2014, 30). These figures suggest that while the market for computer-based software is actually growing, the revenue derived from its sale is decreasing (through a mixture of piracy and a related drop in retail price for individual applications).

This trend has created a significant problem for the music production software industry. The trade body for the sector, the International Music Software Trade Association (IMSTA) includes board members from companies such as Cakewalk, Native Instruments and Steinberg and was specifically set up to address the problem of pirated software through an attempt to 'raise awareness' of piracy issues among music technology consumers. In the organization's initial press release they estimated that a vast majority of music software in use (80 per cent) was pirated and that this figure was, in turn, a much higher percentage than that experienced within the general software industry where the piracy rate is estimated at 36 per cent. In other words, 'for every legal copy of a [music related] software program sold, there are 5 illegal copies in use by potential consumers'.[1] The proliferation of pirated software has been instrumental in encouraging a reduction in the overall price of DAW software, further increasing its accessibility. NAMM's 2014 global round-up of the industry reports Apple's

[1]'Piracy: what do the numbers say', http://www.imsta.org/piracy.php accessed, 20th February 2015.

decision in December 2011 to reduce the price of its Logic DAW to $199 from $499, noting that this put significant 'pricing pressure on competing products' (Challacombe and Block 2014, 27).[2] As a result most companies now offer different versions of their programs at scaled price points, usually starting from around or under $100 (albeit with differences in functionality in the bottom of the range products).

DAWs, deskilling and authenticity

What the above factors point to is an increased accessibility to DAWs and a vastly expanded user base of studio, sampling, sequencing and synthesis technologies. This expansion has been further facilitated by shifts in, and access to, knowledge about music production technology. First, there has been a simplification of the processes of studio recording and composition through the design of DAWs. Secondly, there is much easier access to information about the skill sets and techniques required in their use. The familiarity and ease of use afforded by the drive towards 'intuitiveness' within the design of these applications outlined in Chapter 2 has served to create new skill sets while undermining or even negating others. One of the consequences of this has been the achievability of sounds (and an overall production quality) that would otherwise require considerable technical know-how. For example, the bundling of presets within DAWs provide set schema fixing the variable parameters of plug-in effects and instruments to pre-defined settings. These presets can be inserted into audio or MIDI channels through drop-down menus or drag and drop in order to alter their sound. The naming of these presets offers a quick and logical correlation between a desired sound and its execution. These are often organized around genre, hardware (such as differing types of synthesizer or amplifier), sound source (particular types of microphone positioning, etc.), task (de-essing, cutting or boosting of particular frequencies, etc.), space, or particular well-used effect types (side chain compression, gated reverb, etc.). Logic's bundled

---

[2]Curiously, given the perceived scale of piracy within the sector, software companies have been reluctant to prosecute. A successful lawsuit by the signal processing company Waves against two New York studios who were found to be using cracked versions of their high-end plug-in bundles in 2007 is conspicuous by its singularity. Even when professional producers have admitted to using cracked software, they have remained unprosecuted. For example, in 2009 Jona Bechtolt of the US electronic indie outfit Yacht admitted to using pirated versions of Audio Damage and Ableton. Instead of taking legal action, Audio Damage waged an internet war of words against the producer. This fairly high-profile incident was followed by a subsequent series of similar online 'outings' of well-known producers such as Avicii, Martin Garrix and Steve Aoki who have been taken to task on their use of cracked software in the very public arena of online production tutorial videos and interviews (Tost 2015).

EQ plug-ins, for example, provide presets such as 'rock snare', 'overhead mic', 'tight toms', etc., for drums, 'final mix' dance, ballad, hip-hop, etc., for mastering and 'bass improve', 'add brightness' 'vinyl record improver', etc., for specific audio tasks. The immediacy of these configurations allows users to access fixed professionalized sounds, and generically coherent effects without necessarily knowing their technical properties in terms of acoustic properties, frequency, etc.

Reyes (2010, 330) argues that presets along with other programs such as OpenMix (which provides templates for creating 'professional' sounding mixes) have had the effect of lowering entry barriers to the production process and a 'deskilling' of the types of labour necessary to produce recordings which meet a broadcast or release standard. Reyes's further assertion that this leads to a 'hunt-and-peck music production' (ibid.) (a typing metaphor used to describe untrained and inefficient one-fingered typing) belies a more general set of value judgements surrounding the 'deskilling' of musical practice through technology. Many of the assumptions made about the centrality of DAWs within contemporary music cultures are in fact redolent of long-held discourses that hold technology as a threat. Often these rest upon notions of craft that are embedded within the authenticity paradigms that are characteristic of how creativity is conceptualized within various popular music cultures. For example, Milner quotes a number of long-standing professional rock and pop producers who blame the ubiquity of Pro Tools within the recording industry for producing 'lazy musicians' who produce 'mediocre performances' which are enhanced through editing and manipulation that the program enables (2009, 299). Similarly, digital technologies have been critiqued for eroding 'human' elements of the interaction between musician and instrument in contemporary recordings. This has been positioned as a loss of traditional musical feel (Arditi 2014) or an inability for 'transcendence through flaws' to emerge through nuances and imperfections in traditional music technology (Aho 2009). The rock producer Nick Raskulinecz gives a typical example of this type of perspective in the rock documentary *Sound City*:

> Part of making it in the record business back in the old days was that you could do something and nobody else could do that. Pro Tools has enabled people, any average ordinary person to achieve those results now. ... It's kind of enabled people who have no business being in the music industry to become stars.

Clearly, such reactions from traditional record producers are an understandable response to a perceived encroachment into their field of expertise and an undermining of their own skills base. Nevertheless, their grounding in such loaded concepts of 'craft' and 'feel' links them back to familiar discursive frameworks. Such critiques of DAWs should thus be understood in a historical continuum of popular music authenticity discourses present throughout the age of recorded sound. Auslander (1999), for example, argues that the mediation of music through the recorded work has historically led to an attendant need for aural and visual 'proof' of an artist's musical integrity. From the rock era onwards, live performance served 'to authenticate music as legitimate rock and not synthetic pop' by resolving 'the tension between rock's romantic ideology and the listener's knowledge that the music is produced in a studio' (1999, 79). To a large extent, the fears over digital technologies are an extension of intransigent perspectives relating to various developments in studio technologies along with a long history of the manufacture of acts whose music was in reality the work of behind-the-scenes producers and session musicians.

This is not to say that popular music culture's ongoing discursive relationship to technology is static. Théberge (1997, 211) outlines how the term 'technology' itself tends to be perculiarized and culturally constructed. He notes how as differing technologies have emerged they have been viewed with suspicion within the various discourses of particular popular music cultures. Here, specific types of technology (amplification and electric guitars in rock, synthesizers and sequencing in disco and electropop, etc.) are 'singled out' by the detractors of a musical genre or movement as central to their 'inauthenticity'. Théberge notes that by isolating certain types of new technology as notable (and in many cases by implication 'inauthentic') such assessments render the wider field of technology 'transparent'. In other words, as certain technologies become accepted as a central part of a given genre they are no longer marked out as particularly technological. We can think of this as a naturalization of technology whereby particular types of technological mediation are enculturated to an extent that they become perceived as authentic within particular genre cultures (Thornton 2000). Similarly, Chambers (1992, 194) describes the history of Western popular music itself as being 'the story of a continual appropriation of pop's technology and reproductive capacities'.

While computer-based DAWs remain to some extent emergent in terms of their enculturation as authentic within wider popular music culture, to suggest that they are fundamentally different from previous technologies, or are indicative of a decline in skill in music making, makes little sense. Rather, as the remaining chapters of this book discuss in detail, a new set of authenticities, conventions, skills and virtuosities have emerged that exist concurrently with existing musical and production skills. Certainly, the drive towards intuitiveness that has been a key driver in the development of DAW technologies (outlined in Chapter 2) has meant that such technologies are easier to engage with than, say, a high-end 48-track studio circa 1980. However, this does not mean that either the creative labour invested in contemporary DAWs, or indeed the music made within them, is any less valuable.

The democratization of knowledge

A clear corollary to the DAW's enmeshed relationship to digital and internet culture has been a free dissemination of knowledge relating to their use. This represents a significant shift. As Schmidt Horning (2004) notes, the recording sector has historically been characterized by the accumulation of tacit knowledge acquired through working in the studio where individual recording professionals have been somewhat reluctant to pass on knowledge or have been positively secretive with their methods. The everyday uses of the internet means that not only are the actual technologies needed for sound recording more available, but also information regarding how to use them is very easily accessible. There are countless web forums often with hundreds of thousands of users where users discuss DAWs, plug-ins and instruments, share information about techniques and provide solutions to given production problems, often in a very detailed manner. Similarly, video sites provide a major resource in this respect. For example, searches for production tutorials on the video content site YouTube produce a staggering number of results: 492,000 for Pro Tools, 292,000 for Logic Pro, 232,000 for Ableton and 427,000 for FL Studio. This allows information to be presented in an accessible way as the video format facilitates clear instruction with most tutorials consisting of walk-throughs of specific tasks using the DAW's graphical user interface to illustrate exactly how a particular task is undertaken.

These online spaces function as virtual communities of practice (Wenger 1998) or 'knowledge communities' (Salavuo 2006) whereby learning is a social process in which interested parties collectively find solutions to given problems and progress their individual and collective knowledge. Such spaces are now a major way in which musical knowledge is transmitted. As various studies across instrumentation and genre have shown, the ability to transcend time and space helps facilitate self-directed lifelong music learning outside of formalized educational structures for large groups of people (Rudolf and Frankel 2009; Waldron 2009; Kruse 2012).

Interestingly, a knock-on effect of the emergence of these virtual spaces has been a loosening of the protection of information relating to professional practice. There are a number of web-based channels that freely disseminate 'insider' knowledge about production techniques that in a previous era would have been closely guarded. Perhaps the most well known of these has been the weekly web TV show *Pensado's Place* which approaches a million views per month. Presented by the mix engineer Dave Pensado it takes a talk show format in which guests (generally producers and engineers from the upper echelons of the recording industry) talk through various aspects of the creative process. The show also includes detailed tutorial sessions on specific aspects of production from creating drum loops to compression, mixing and mic techniques. Pensado acknowledges that the establishment of the show was in reaction to the high demand for online instruction. Most individuals with access to studio equipment are now self-taught and the changing roles facilitated by the predominance of DAWs mean that the knowledge germane to music production is shifting:

> Everything is tied to the digital space. At one point, you had to be a millionaire to access studio time. Now you can access the same equipment for a few thousand dollars. If a person can have access so inexpensively, he or she can also become self-taught. ... I mix records for a living, but I don't know how much longer that will be a professional option. Songwriters are recording their own demos, adding plug-ins and mixing. You can't spend a whole year teaching someone how to EQ vocals anymore. ... [Educators] have to understand this new generation's culture and mores. You can't attract them with old methods. (Mitchell 2013, 23)

In a sense, Pensado's comments get to the crux of the effects of changing technologies: there is a high propensity towards the autodidactic, there is a conflation of creative and technical roles within the recording process, and presets and 'out-of-the-box' ease of use have rendered certain knowledges as unimportant. These factors are becoming increasingly significant in the trajectory of a generation of younger producers entering the professional field. For instance, Carlo 'Illangelo' Montagnese, one of the emergent high-level pop producers of 2015, commented that his way into production and his subsequent production style was directly resultant from his engagement with internet technologies:

> I asked ... [my brother] to show me how to sequence on an old Roland and he told me that I could find all the info online. So I spent a lot of time using online resources. ... Someone told me that Cubase was the professional software to use so I downloaded a cracked copy and learnt my way up from there. I was doing everything, as I still do today: writing, playing, engineering, mixing, producing mastering. (Tingen 2015)

What, then, are we to make of this recently emerged landscape of digital production in terms of democratization? While figures relating to cracked software are somewhat imprecise and to a certain extent guarded by the software industry, it is possible to draw some general conclusions. First, the availability of various types of software that function as studio and/or compositional tools has vastly opened up in comparison to when the amateur recording and electronic production market first emerged in the 1980s. The availability of free software, a reduction in price of legal applications and their integration into the quotidian environment of the personal computer have significantly shifted barriers to entry. Secondly, knowledge around recording and production techniques no longer requires the immersion in professional networks for the acquisition of technical knowledge in the way that it had done throughout the twentieth century. In light of this situation, it is reasonable to conclude that the amount of people making music in this way has seen an enormous increase.

Democratization distribution and promotion

In one way the vastly increased number of individuals involved in production can be read as a democratization of practice. However, this does not automatically represent democratization in a structural sense. The fact that there are more people involved in production does not necessarily lead to changes in the political economy of music making or to the structures and power relations of its institutions. Nevertheless, the technological changes outlined here are often infused with the utopian/disruptive discourses outlined at the start of this chapter. In particular, web-based communications have been widely read as constituting a democratization of distribution facilitated by the so-called symmetrical nature of internet technologies (where there is an equality between whose message can be heard on a global scale) (Bruns 2010, 24). The technical ability to distribute information and digitized cultural products at little cost to a potentially global audience has been at the heart of the celebratory reception of digital technologies. The democratization of distribution argument suggests an open field in which audiences are available easily and readily for anyone who produces music. Online platforms such as Soundcloud, YouTube, iTunes and Bandcamp act as global aggregators, central nodal points through which cultural producers can reach worldwide audiences. The emergence of these platforms are seen to have had profound effects and are often read as inherent to the social and technological circumstances of the post-digital context. For example, in placing the emergence of DAWs into the historical context of attempted democratization within popular music culture Prior argues that 'the difference today is one of global reach, speed, ease of use, and absolute scale. One might even suggest that the DIY ethic so cherished by punk rockers is no longer an activist ideology, but a systematic, structural condition of the production of music itself' (Prior 2010, 404).

However, such a straightforward corollary from digitization to democratization needs to be problematized. We should not see the field of musical production overall as a singular open field that has been democratized by digitization. Rather, we should perhaps understand channels of distribution and consumption as a multifarious collection of different fields ranging from social networks based around amateur music making, micro and DIY production, independent production based around different genre cultures and the mainstream recording industry. Each in their own way has been affected and changed by digitization and has responded and adapted to changes in technology. Nevertheless, these differing fields of production operate in an autonomous or relatively autonomous manner, sometimes overlapping, but largely existing according to their own logics and internal networks.

Amateur production

In terms of amateur music making, material is often distributed and promoted via message boards, the social networking aspect of uploading sites such as Soundcloud, Reverbnation, Soundclick, and file-sharing networks (Whelan 2006) as well as through video sites such as YouTube. Clearly, such platforms are designed to facilitate communication and their very nature allows for individual producers work to be heard by large numbers of listeners outside of their immediate social and geographical location. Much of the production facilitated and shared through such sites is by its nature resolutely amateur and non-commercial. In Salavuo's empirical study of online music making communities for instance, respondents framed their involvement within such networks in terms of affirmation, collaboration, sociality and the opportunity to get feedback on their work. Here, just 10 per cent named the 'economic benefits, such as obtaining a recording contract or becoming famous and making money through offering songs, as important factors for their participation' (Salavuo 2006, 265).

Nevertheless, it is undeniable that certain types of amateur production have had a significant cultural impact. For instance, the mash up and remix culture that quickly emerged in the wake of Web 2.0 in the early-to-mid-2000s was largely an amateur phenomenon facilitated by freely available production software and the ability of sound and video platforms which enabled tracks to go viral through social media. As with the warez scene there is something about these practices that is clearly reflective of wider internet discourse and culture with its open source ethos (the term 'mash up' itself also simultaneously refers to patterns in computer programming and application development). The recontextualizing and reworking of existing texts which is central to both remix and mash ups facilitate an engagement with musical texts that reorientate the producer/consumer binary and call into question the stability of texts produced by large media conglomerations. Bruns (2010) situates these practices as 'produsage' where media content is realized not through 'a conventional production process that is orchestrated and coordinated from central office and proceeds in a more or less orderly fashion to its intended conclusion (the completion of the finished product)'. Instead, content creation is undertaken by 'produsers', a significant segment of the media audience who are both user and producer, and 'constitutes an always ongoing, never finished process of content development and redevelopment which on occasion may fork to explore a number of different potential directions for further development at one and the same time' (Bruns 2010, 26). This blurring of roles between producer and consumer and the fundamental practices of mash up culture have been seen to undermine existing power relationships of cultural production and destabilize the underlying logic of ownership and exploitation of cultural works that have been central to the existence of large media companies.

The mash up's intersection of free distribution beyond the realms of big media and its inherent critique of intellectual property legislation has meant that it has been predominantly framed within both scholarship (McLeod 2005; Serazio 2008; Shiga 2007) and popular media as an example of digital technology's capacity to build 'democracy while dismantling capitalism' (Vallee 2013, 77). Perhaps the most notable mash up artist is the US artist Girl Talk (Gregg Gillis), whose work and public persona have been reflective of this position. Resolute in his engagement with the politics of intellectual property and his commitment to the original aesthetics of mash up culture, his five collage-style albums were released under a Creative Commons licence and made available via the Illegal Art website where downloaders could pay whatever fee they saw fit. On the back of these releases, Girl Talk has managed to carve a lucrative decade-long career on the US club and festival circuit and has been seen as something of a symbol of a new type of cultural production facilitated by remix culture (Cohen 2014).

However, despite the undeniable significance of mash up and remix culture as a cultural phenomenon, to attribute any long-lasting structural change upon the music industries to it would be simply erroneous. While mash up culture continues to exist, facilitated through a number of key web portals, large media corporations have responded to it in two central ways: on the one hand, controlling and litigating against any copyright infringement considered significant in terms of threat to their business model, and on the other, exploiting elements of mash up culture for their own ends. Probably the most watched video to emerge from this culture was the 2011 live mash up 'Pop Culture' by Madeon, a then sixteen-year-old French production prodigy who garnered over 30 million hits on YouTube with a track using thirty-nine samples from pop hits, mixed in Ableton Live and controlled live via a Novation Launchpad MIDI controller. After the success of the video, Madeon almost immediately signed to Sony and began to undertake more traditional production and remix work, scoring a number one album in the US Dance Chart in 2015. Similarly, most of the other figures behind notable mash up viral successes have subsequently gone on to shift their practice towards more traditional production models. Early mash up artists such as Richard X and Mark Vidler have made the leap into more traditional record production or have international DJ circuit. In this respect, Girl Talk remains something of an exception in a set of relationships characterized by co-optation and adaptation of practice.

Processes of co-optation can be seen in other facets of the relationship between Web 2.0 and music production. The dominant uploading app Soundcloud provides a good example. Soundcloud functions for many music makers as a social network where producers and listeners can interact with each other, commenting on individual tracks often with a view to collaboration, exchange and remixing possibilities. Remixers and DJs (whose work often utilized existing copyrighted material) formed a core part of its constituency in the early period of its development and were crucial in its transition from start-up to dominant platform. On a basic level Soundcloud provides a platform for an individual's music to be heard by a worldwide audience. Since its launch in 2008 the Soundcloud platform has been promoted and understood as a key way of gaining an audience and gaining access to distribution and promotion channels. Building up a strong base of followers and gaining significant amounts of plays for individual tracks (especially in EDM which has been a dominant genre in Soundcloud's emergence) have become central tools in new artists' attempts to launch their careers. While there are a number of examples where tracks have become commercially and critically successful through an initial launch on Soundcloud there are a number of caveats that should be taken into account. Due to the lack of monetization facility for independent artists, any ongoing success or financial sustainability remains dependent on validation and sponsorship from, and interaction with, established cultural intermediaries such as record company staff and music journalists and established bloggers. The use of the platform in this way can be read as providing something of an A&R shop window which confirms binaries of 'signed' and 'unsigned' artists and replicating the existing power relationships of popular music practice in which unknown artists attempt to gain the attention of, and subsequent sponsorship by, key gatekeepers who can provide access to the necessary channels for an artist to gain monetary or critical success. In general, the platform's model does not allow for significant monetization in itself for an artist unsponsored by a larger music company to be self-sustaining. The way in which artists or labels can make money from the platform is highly geared towards those who are already established or have the necessary connections within the recording industry. The platform's On Soundcloud program launched in 2014 provides a way for 'Premier Partners' to monetize their tracks through a share in advertising revenue tracked through plays. The program was launched as invite only and included just 100 partners in the first six months of its operation, all of which (including Warners, Universal, a number of the larger EDM independents along with highly popular spoken word and comedy content) were existing major players in their respective fields (Houghton 2015). The connection between the platform and the large music companies was further tightened in early 2016 when Soundcloud struck a licensing deal with Universal for all of their content. The closer involvement with the majors by Soundcloud was in itself a natural result of the companies' need to monetize its business model. As part of this relationship the company became increasingly strict in enforcing intellectual property rights on the behalf of the major labels, often alienating the very DJs and remixers that had formed a core part of the user group and had been instrumental in its growth with cease and desist notices and the deletion of material.

Throughout the history of the recording industry it is possible to trace repeated examples whereby new technologies are identified, exploited and co-opted by large music companies. The instance of Soundcloud has parallels with the way that the major labels reacted to independent project studio production in the 1980s. The success of independent producers in creating commercially successful electronic music was hailed as a stride towards greater independence from standard corporate structures within the recording industry. However, in addressing the issue of technology and democratization in this respect Hesmondhalgh (1997b) is circumspect. In a discussion of the UK dance music industry, he argues that although that by the mid-1990s the 'bedroom studio' had been proclaimed as the key site of democratization by dance music's protagonists and apologists, the actual effect of new technologies did not necessarily constitute fundamental change. For Hesmondhalgh, the fact that records could now be produced in small-scale project studios was of little relevance to wider structural change in the music industry as they were still primarily signed, licensed and distributed by large corporations if they were to access the market to any great significance. In effect, although the intervening two decades have seen rapid technological development, the structural inertia resultant of the dominance of large corporations in the era of the physical production and distribution of recording to a large extent remains.

This is something that has been noted across differing sectors of cultural production in the wake of digitization. The case studies included in Buckingham and Willet's (2009) project on digital video cultures suggest that we should be cautious about attributing 'revolutionary' effects to contemporary amateur and prosumer uses of creative digital technologies. Across divergent cultural industries such as journalism and pornography they argue that large media corporations have been 'very effective in recuperating any potential challenge represented by amateur producers, and in employing "user-generated content" to their own advantage' (2009, 231). This is certainly true of the recording industry, which, despite the undoubtedly negative effects upon music sales resulting from illegal downloads and the significantly lowered revenue to be garnered from streaming services, has successfully exploited Web 2.0 in terms of mediation and marketing. Throughout the 2000s web marketing has become increasingly important to the large-scale, well-funded marketing campaigns of the major recording companies whereby social network campaigns and viral marketing have come to the forefront. As Wikström (2009, 162) notes 'through the growing importance of social networks, these kinds of uncontrolled processes [viral, consumer led success stories] are no longer merely random flukes, but are ... becoming more or less the norm of music promotion in the digital age'.

The marketing strength of viral campaigns is that they give the appearance that success has been achieved without the overt control of large media corporations by a natural groundswell of demand from individuals rather than being orchestrated through the more familiar systems of promotion. The historically embedded discourses of authenticity, sincerity and a concern with the contradictions of art and commerce that have been central to popular music culture since the 1950s are, in a sense, replicated through the ways in which the idea and appearance of democratization has been co-opted by the music industry. As I have noted elsewhere, the idea of a democratized media has in itself become a central part of the discursive practices of contemporary marketing campaigns. For example, acts as diverse as the multimillion-selling UK indie act the Arctic Monkeys, British pop star Lilly Allen and US indie band Clap Your Hands Say Yeah were all promoted having an organic, grass-roots web-based fan-centred rise to prominence. The narrative was that their talent had been recognized directly by audiences which in turn had driven their rise to fame, conveniently leaving out the fact each had very experienced management, recording companies or marketing teams behind them (Strachan 2007). Similarly, McLean et al. (2010) and Salvato (2009) examine how the UK pop artist Sandi Thom and US singer-songwriter Marié Digby, respectively, were subject to a very expensive and highly coordinated web launches which actively constructed them as DIY artists; Thom through the use of misleading webcasts purportedly broadcast from the singer's basement and Digby through a series of apparently self-filmed YouTube videos in which she played acoustic covers of popular chart hits.

We can read these examples of record companies appropriating the means, aesthetics and strategies of democratized new media as one way in which large media companies have attempted to reposition themselves in the light of technological change. In tandem with attempts to control copyright infringement through legal action there has been a rearguard action to monetize recent trends in the dissemination of music at all levels. After a period of considerable instability the major music companies had to take a concerted attempt to re-establish control over the market in terms of relevant emergent revenue streams. A significant recent strategy has been to buy up segments of relevant digital entertainment start-ups. The remaining 'big three' music companies Warner, Universal and Sony have all gained substantial equity stakes in streaming services such as Spotify and Rdio, the song recognition service Shazam and Soundcloud. As O'Malley, Greenburg and Messitte (2015) have noted this has been achieved through the usage of the major labels' existing artists as leverage when striking licensing deals often with the clause of leaving open the option to buy stakes in the relevant company. These investments in delivery services are in keeping with the strategies of vertical integration that have been characteristic of the recording industry since the 1920s whereby through multiple mergers, buyouts and joint ventures, large media companies sought to control all aspects of the production chain from recording and publishing through to manufacture, distribution and sales. The negotiating power of the major music companies is further illustrated by the way in which revenue is divided under the terms of arrangements with streaming and downloading services. For example, findings by the market research company Ernst & Young reveal that percentages of streaming revenue from the services Spotify and Deezer were heavily weighted in favour of labels at the expense of artists. The report found that 73 per cent of payouts derived from the premium subscriptions of streaming services were paid out to labels, with publishing rights accounting for 16 per cent and artists 11 per cent (see Ingham 2015).

This is not to say that independent artists or self-produced recordings by individuals without the help of a record label cannot break through into the mainstream. Rather, that these instances remain somewhat rare as through processes of consolidation, investment in digital start-ups and changing promotional strategies and priorities, the oligopoly of large media corporations is maintained and policed. Furthermore, by attempting to control (and gain a stake in) the platforms that host and aggregate all types of music (from major label through independent to amateur) big media is clearly seeking to consolidate its position in terms of the exploitation of intellectual property rights that has been partially eroded through digitization while simultaneously opening up further revenue streams. Digitization has enabled numerous ways of distributing music and the major music companies are finding ways to try and maximize the revenue that can be accrued from each. As McClean et al. (2010, 1373) conclude, despite Web 2.0's capacity to make more music more widely available, 'ultimate power remains with the major labels, mass media and broadcasting rather than independent "narrowcast" and DIY artistry. Music continues to be commoditized and fans continue to be constructed as "consumers".' This is backed up by quantitative survey work such as Leenders et al. whose research did not find 'strong support for a disintermediation effect for most types of artists in relation to their market access' (2015, 1813). They go on to suggest that the group that benefitted most from a range of new media were emerging artists signed to large recording companies and conclude that 'record labels seem firmly in control of the access to the (global) mass media and they are also using new media quite effectively in their bundling strategies' (ibid.) This idea of Web 2.0 campaigns being part of a bundle of effective strategies employed by the large music companies is, I think, crucial in understanding the realities of how new media has shifted the landscape of popular music marketing and promotion. Techniques such as viral marketing, co-creational engagement and social media campaigns tend to be part of highly orchestrated, carefully timed, multipronged campaigns. In most cases, for a campaign to successfully penetrate the market, Web 2.0 strategies are employed alongside traditional media exposure via TV and radio campaigns, the use of key print- and web-based journalism, touring, cross-promotion, licensing, etc. In turn, it is clear that the multi-faceted nature of contemporary promotional conventions requires significant financial resources and trusted professional relationships across a complex web of institutions and cultural intermediaries. The control of finance, professional connections and institutional relationships are crucial accessing the market in meaningful ways, and are thus central in the ongoing maintenance of the power structures within the field of production. Given the diminishing importance of manufacture and distribution to the recording industry, marketing and promotion occupy a position of increasing centrality as a core function of what major music companies do, and how they control access to the market. As Jones (2012, 185) notes 'taste making has not disappeared' and market entry is dependent on organizations which have the necessary strategies and institutional clout to convince media outlets 'to alert their audiences, as target markets, to new releases'.

The independent sector

The continuation of an existing hegemony can also be observed in relationship to the independent sector. Again, for techno-utopians digitization has been seen as something of a levelling of the playing field for independent record labels and cultural producers. For example, McLeod's assessment of these relationships in the mid-2000s argued that internet technologies and the crisis in the majors caused by file-sharing offered clear opportunities for the independent sector. For McLeod 'internet distribution' provided a clear challenge to the oligopoly of the majors and the emergence of new technologies meant that 'independent artists and the owners of indie and micro-labels are able to turn into an advantage what used to be a liability: lack of major label connections and relative geographic isolation' (2005a, 527). However, subsequent research, both from a statistical and qualitative methodological perspective, has painted a much less optimistic picture of the implications of digitization upon the independent sector. Rogers's (2013, 144–9) empirical research with music industry professionals suggests that while the internet emerged with clear potential for the independent sector, it has worked best for established record companies and to a certain extent, for very narrow niche markets. He quotes interview material from one small label MD as being typical of the experiences of the independent sector, commenting that 'those who bought into the idea of the internet taking the major companies out of the picture were wrong from the start. That has not happened and is not going to happen' (Rogers 2013, 147). This position is backed up by Arditi (2014, 411–12) who points to sales statistics that indicate that the major labels have maintained their oligopoly and have actually increased their market share in the post-digitization environment. In addition, most independent labels do not have the necessary leverage to negotiate with large new media corporations on equal terms with the majors and are thus are subject to inferior licensing deals. As Sisario (2014) notes, 'small labels complain that consolidation by the major record companies has left them squeezed in negotiations with the online music services that now account for a majority of their revenue'. For example, a document prepared for a US government judiciary subcommittee by Darius Van Arman of the A2IM independent trade body argued that through practices such as the stipulation of large advances from streaming services (which in turn leaves 'no recourse left to the digital service other than to heavily discount what they can offer as compensation to independent rights holders' (Van Arman 2015, 5)) and the control and unfair distribution of breakages[3] 'the three major recording companies have become proficient at extracting a disproportionate share of copyright-related revenue from the marketplace' (ibid.). Even a relatively optimistic empirical account such as Hracs (2012, 458) concludes that real term income derived from music for independent performers has significantly decreased in the first decade of the millennium.

What such accounts suggest, then, is a continuing structural dominance in which the independent sector faces a similar set of barriers and challenges that have always existed within the recording industry. In a realistic analysis of the effects of digitization Hesmondhalgh (2007, 261) gives a succinct assessment of the internet's impact upon the social relations of cultural production. He points out that it is undeniable that the internet has facilitated enormous amounts of small-scale cultural activity, provided new channels of communication and has been utilized in innovative modes critique and activism. However, this must be tempered by the fact that these trends have done little to alter the enormous 'concentrations of power' (ibid.) within the cultural industries. He concludes that instead of a deeply transformative democratization, these developments constitute a series of 'disturbances' to the political economy.

Overall, the picture of the differing levels of cultural production outlined in this chapter so far points to convergent stage of digitization from the mid-1990s as constituting a period of significant change. However, the processes of consolidation and adaptation within the major music companies have illustrated that to characterize these changes as a root and branch democratization is highly problematic. Yet there is no doubt that there are certain elements of the production process upon which the effect of convergent digitization is real and profound. While these 'disturbances' can be seen to have little impact upon the power relationships of the media industries, it is nevertheless an undeniable fact that recent trends in digitization have had a clear effect on significant elements of popular music practice. While the

[3]As major rights owners, major labels have been able to negotiate substantial advances recoupable for streaming services that are non-refundable. As part of this system, royalties owed to the label through the streaming of their tracks are deducted from the advance. At the end of a given collection period if the advance exceeds the amount that the digital platform owed the label for content used, the record company keeps the remainder of the advance. Instead of redistributing these 'breakages' among artists, it has been pointed out that there has been an extreme lack of transparency as to where these unaccounted for profits end up (see Cooke 2015).

macro structures relating to music and media have adapted and shifted in order to maintain their structural dominance, the micro modalities of creative practice and cultural production are clearly transformed in other ways.

The continuing dominance of the major music companies has meant that to a large extent the independent sector has continued to exist according to a set of structures and conventions that have existed for a number of years. On one hand, independent labels and musicians are situated within a complex web of relationships between majors and minors (Negus 1992, 18) through which industrial practices negotiated alliances, licensing deals, joint ventures and buyouts. On the other, many cultures have constituted separate fields of production autonomous from, and often at odds with, the mainstream market and its industrial networks (Kruse 2003; Hibbett 2005; Strachan 2007; Moore 2007). While both of these modes of independence remain relatively intact, convergent digitization has been nonetheless instrumental in a series of aesthetic developments and has affected the shape and nature of creative labour within the relative fields. Partially, this has to do with the emergence of new musical styles and genres and the manner in which they are spread and facilitated. For example, Chapter 5 argues that the social and cultural conditions of digitalization have been central in the emergence of a distinct set of cyber genres closely linked to independent and DIY production in the past decade and a half. The chapter argues that the genres are both reflective of our relationships technology in an aesthetic sense and ultimately reliant on web-based communication for their emergence as geographical spread. Perhaps the clearest shift, however, is in terms of division of labour, the shifting roles of creative agents within the production process and the amount of agency that individuals can now exercise in the creative process. The emergence of the DAW has been central to these developments. Hugill (2008, 188–9) sees the advent of the DAW as both a logical progression from the idea of the studio as a compositional and creative space that had been apparent in various strands in musical culture since the 1950s, and a central driving factor in the emergence of what he calls the 'digital musician'. The impact of virtual studio technology and the integration of various roles by electronic musicians more generally have led to unwieldy (yet understandable) descriptive terms such as 'writer-producer-engineer-performer' (Moorefield 2005, 98). Whatever the terminology it is clear that cross a variety of genres musicians are no longer 'just' musicians. They engage in a variety of practices that have more commonly been associated with recording studio staff such as engineer or producers. To a large extent across a variety of contexts, the terms of being a musician have shifted as individuals accumulate knowledge that remained somewhat obscured within the more defined division of labour that was engendered when recording studios constituted the only viable site for 'professional standard' recordings.

The changes in recording technologies have thus had implications for certain sectors of musical production in terms of the roles of cultural intermediaries and their control over the creative process. Musicians releasing music through independent labels may actually be in close control of not only the composition of their own music but all of the technical aspects of the final recording (engineering, editing, mixing, etc.) up to the mastering process. As Ryan and Hughes (2006) point out, whereas previous recording practices had necessitated creative collaboration with differing key decision makers, DAWs have given artists the ability to produce and distribute a sizeable quantity of material on their own without the intervention and scrutiny of producers, A&R personnel and managers. While labels continue to act in an intermediary role in terms of the selection of material marketing and may provide feedback and suggestions for individual tracks or the overall direction of a particular project, there is little doubt that the musician producer is afforded increased autonomy facilitated by the possibilities of digital technology.

In addition, the emergence of new technologies more broadly has led to the shifting of, and elision between, roles within the commercial practices of music. Individuals within small-scale cultural fields of production often fulfil a variety of roles. As Hracs (2012, 458) notes, digital technologies have meant that independent musicians have taken on increasing amounts of entrepreneurial and administrative tasks as well as undertaking more conventionally creative activities. For example, within various strains of independent electronic music, many musicians are actually involved in running labels themselves. Many of the artist/producers interviewed during the researching of this book have been involved in the running of successful independent labels. Shackleton jointly ran the Skull Disco imprint, Matthew Herbert has been involved with Accidental Records for over a decade, Geiom ran the imprint Berkane Sol, Robin Saville of Isan runs Arable, Carsten Nicoli is co-owner of Raster Noton while Donnacha Costello has had success with a number of unlabelled self-releases. While all of these artists have primarily released their material through a variety of other labels run by other people, they are illustrative of how normative it has become for creative individuals to have an integrated approach to the cultural production process. In this context, running a label is often part of a wider portfolio of activities that may supplement other income and maintains an artist's profile within electronic music networks.

In terms of how the cultural production of music has been traditionally organized we can think of such individuals as integrated small-scale cultural practitioners. The culture of production approach that has been widely applied to the music industries has tended to view the production and distribution of recordings as taking place within a connected network (Peterson 2001; Zolberg 1990; Negus 1992). Each point within this network is involved in a particular task through which artistic works/cultural products are sponsored, created, promoted, validated and distributed. In turn, the works themselves are necessarily shaped by the structures of the institutions in which they are produced and distributed. The integration (vertical and horizontal) and conglomeration of these elements by large corporations has been an historical feature of the recording industry. Integration has been central in the maintenance of control and power within the cultural production process. The small-scale nature of cultural production alongside increased connectivity facilitated by Web 2.0 produces an economy of scale that enables the integration of roles (and thus power and control) and (at least) the possibility of financial sustainability. The centrality of integrated small-scale cultural practitioners within underground cultures and sub-genres of popular music such as electronica, lo-fi, glitch, etc., affords a level of creative autonomy at odds with the more stratified structures of the mainstream recording industry. This does not mean, however, that these cultural producers are free from institutional constraints. Rather, they are enmeshed in a rather more defuse, but nonetheless powerful, web of power relationships. They are reliant on the structures and networks pertaining to their particular genre cultures and require validation and sponsorship by cultural intermediaries pertaining to that culture. The financial viability and sustainability of artists' activities are dependent upon their connections with taste makers, including web media, blog writers, record labels, promoters, festival organizers, etc., who form the central nodes controlling distribution and promotion within such genre cultures.

DAWs and mainstream record production

These changes in creative practice do not just sit at the level of independent, DIY or niche music production as they can be identified within music production which reaches the widest possible audience. DAW technologies are increasingly at the heart of mainstream record production. Since Ricky Martin's 1999 single 'La Vida Loca' became the first global hit record to be entirely recorded and mixed in Pro Tools, DAWs have become central to a variety of recording practices across differing popular music genres. In the period since, Pro Tools has become something of an industry standard production platform that is at the heart of high-end studios across the recording industry. For example, in Théberge's statistical analysis of the US and Canadian studio sectors he concludes that by 2007 'virtually every studio seemed to offer some kind of Pro Tools compatibility' (2012, 85). Further, the technology used to produce internationally successful recordings is, in many cases, the same, and is widely available to anyone with a personal computer and a relatively small financial outlay. This, I think, is crucial in understanding the implications of the rise of DAWs upon popular music practice as it has the material effect upon modes of creative labour and economies of scale within the recording industry.

In order to examine the scale and scope of these changes, the final section of the chapter will provide an in-depth analysis of the UK top 30, bestselling singles in a randomly chosen week in 2015 (18th January) in order to assess the types of platforms used and the shifting nature of the studio in the contemporary recording sector. Taking the chart as starting point, I researched the production processes of each track. Through an examination of the production credits for each of the entries on the charts combined with publicly available interview material and equipment lists from recording studios, a full picture was built of how each of the entries was written, recorded and produced. This information is presented in Figure 1.1 which breaks down the information into track, chart position, artist, producer(s), studio and recording platform. The chart contained nineteen entries from British or Irish artists, three from artists from other European countries and eight which included US-based artists. The chart was dominated by the major music companies with twenty-eight of the artists being signed to major labels or their subsidiaries, while the two others were released on established independent dance labels based in the UK. All entries (bar UK TV talent show winner Ben Haenow) had some level of international chart success outside of the UK and Ireland with twenty of the tracks also making the US Billboard Hot 100 and a further three making the top 10 of the US Billboard Dance Chart.

The most obvious and immediate aspect of the breakdown of recording context here is that it illustrates the complete enculturation of the DAW into contemporary recording practice. Every entry utilized some form of DAW in its production. It is perhaps significant that the only example of a tape-based studio to be found in the entries was tracking undertaken at Dunham Studios, Brooklyn, which has made its name through the production of self-consciously retro productions.[4] Secondly, the type of record that dominates the chart illustrates the predominance of electronic music and electronic production styles in the contemporary pop landscape. The chart contained singles from differing types of mainstream pop from singer-songwriters through to various types of electronic dance music producers through to global pop superstars but is predominantly characterized by the use of electronic drum beats, synthesizers and sampling technologies.

Another striking feature of the chart is the diversity of applications used in the production of the tracks it contains. While Pro Tools is clearly dominant, the fact that twelve of the entries were created using Logic and a further three used FL Studio, Cubase and Ableton Live suggests that there is no one 'industry-standard' DAW. The fact that many entries do not use Pro Tools also reveals diversity in the type of studio that is being utilized at this level of the popular music industry. The classification of a 'professional' studio as consisting of a control room with a large amount of outboard

---

[4]The studio is affiliated to the Daptone label, which has in the 2010s scored some critical and commercial success with artists (such as Charles Bradley and Karen Jones) who draw specifically on the styles and production techniques of 1960s Soul music. The studio is perhaps best known as the location of much of the tracking for Amy Winehouse's *Back to Black* in 2006.

Track	Chart Position	Artist	Producer(s)	Studio	Platform
Uptown Funk (feat. Bruno Mars)	1	Mark Ronson	Bruno Mars, Jeff Bhasker, Mark Ronson	Dunham, Brooklyn Zelig, London (artist's own studio), Enormous, Los Angeles (producer-owned studio)	Analogue tape, Pro Tools
Wish You Were Mine	2	Philip George	Philip George	Artist's own project studio, Nottingham	Logic
Take Me To Church	3	Hozier	Hozier, Rob Kirwan	Artist's own project studio, Wicklow and Exchequer, Dublin (producer-owned studio)	Logic, Pro Tools
Up (feat. Demi Lovato)	4	Olly Murs	Cutfather, Daniel Davidsen, Peter Wallevik	Air Edel, London	Logic, Pro Tools
Thinking Out Loud	5	Ed Sheerin	Jake Gosling	Sticky, Surrey (producer-owned studio)	Cubase, Ableton Live
Blank Space	6	Taylor Swift	Max Martin, Shellback	Conway Studios, Los Angeles	Pro Tools
Heroes (We Could Be) (feat. Tove Lo)	7	Alesso	Alesso	Artist's own project studio, Stockholm	Logic
Something I Need	8	Ben Haenow	John Ryan	Sarm, London	Pro Tools
Promesses (feat. Kaleem Taylor)	9	Tchami	Tchami	Producer's own studio	PreSonus Studio One
Like I Can	10	Sam Smith	Jimmy Napes, Steve Fitzmaurice	RAK, London Producer's own writing studio, London	Pro Tools
All About That Bass	11	Meghan Trainor	Kevin Kadish	Carriage House, Connecticut. (producer's own studio)	Pro Tools
Real Love	12	Clean Bandit and Jess Glynne	Clean Bandit	Artists' own project studio, London	Ableton Live

FIGURE 1.1 *UK Top Thirty Tracks 18 January 2015.*

Elastic Heart (feat. The Weeknd & Diplo)	13	Sia	Greg Kurstin, Diplo	Echo, Los Angeles (producer's own studio)	Logic
Outside (feat. Ellie Goulding)	14	Calvin Harris	Calvin Harris	Fly Eye, London (artist's own studio)	Logic
I Loved You (feat. Melissa Steel)	15	Blonde	Adam Englefield and Jacob Manson	Artist's own project studio, Bristol	Logic
The Nights	16	Avicii	Avicii	Artist's own project studio, Los Angeles	FL Studio
Budapest	17	George Ezra	Cameron Blackwood	Voltaire Road, London (producer-owned studio)	Pro Tools
Night Changes	18	One Direction	Julian Bunetta and John Ryan	Wendyhouse, London. Various mobile locations.	Logic
Chandelier	19	Sia	Greg Kurstin	Echo, California (producer's own studio)	Logic
Wrapped Up (feat. Travie McCoy)	20	Olly Murs	Steve Robson	Northern Sky, London (producer's own studio)	Pro Tools
I'm not the Only One	21	Sam Smith	Jimmy Napes, Steve Fitzmaurice	RAK, London Producer's own writing studio	Pro Tools
Dangerous	22	David Guetta feat Sam Martin	David Guetta, Jason Evigan, Sam Martin	Power Sound, Amsterdam, Piano Music Amsterdam	Pro Tools, Logic
Bang Bang	23	Jessie J/ Grande/ Mina J	Max Martin, Rickard Göransson	Maratone, Stockholm (producer-owned studio) Metropolis Studios, London, Conway Studios, Los Angeles Glenwood Place, California	Pro Tools
Stay With Me	24	Sam Smith	Jimmy Napes, Steve Fitzmaurice	RAK, London, Producer's own writing studio	Pro Tools
Blame it on Me	25	George Ezra	Cameron Blackwood	Voltaire Road (producer-owned studio)	Pro Tools

FIGURE 1.1 (CONTINUED) *UK Top Thirty Tracks 18 January 2015.*

Shake it Off	26	Taylor Swift	Max Martin, Shellback	Conway Studios, Los Angeles	Pro Tools
These Days	27	Take That	Greg Kurstin	Echo, Los Angeles (producer's own studio)	Logic
Cool Kids	28	Echosmith	Mike Elizondo	Lightning Sound, California	Pro Tools
Don't	29	Ed Sheerin	Rick Rubin, Benny Blanco	Producer's own project studio, Shangri-La, California (producer-owned studio)	Pro Tools
Steal My Girl	30	One Direction	Julian Bunetta, Pär Westerlund and John Ryan	Wendyhouse, London, various mobile locations.	Logic

FIGURE 1.1 (CONTINUED) *UK Top Thirty Tracks 18 January 2015.*

gear and a large mixing desk alongside a live room for the recording of acoustic instruments is problematized by the way in which many of the tracks were made. The chart includes a number of recordings made in the respective producer's own DAW-based project studio often entirely recorded and mixed 'in the box'. The top 20 included the work of: 21-year-old Philip George who produced the number 2 single 'Wish You Were Mine' in his Logic-based bedroom studio in Nottingham and had amassed 2,000,000 plays of the track on his Soundcloud page before gaining a licensing deal for the track from UK independent label 3 Beat (Wilson 2015); Avicii, a superstar Swedish producer who has scored a series of international hits from a project studio based around FL Studio; Alesso, another Swede who despite working with global stars such as Calvin Harris and David Guetta, composes and mixes entirely 'in the box' using Logic Pro (Barker 2012); Tchami, a French producer/DJ who is deliberately evasive about both his persona and studio set-up (but who from his instagram postings clearly uses the PreSonus Studio One DAW and VST plug-ins in a home studio); Clean Bandit, a UK electropop band who record in a converted portacabin in Kilburn, London using Ableton Live, again entirely 'in the box' (Barker 2013). Other entries from production teams who primarily use Logic in their own studios included LA-based Greg Kurstin who produced Sia's two entries in the chart and Take That's 'These Day's', the Danish team of Cutfather, Davidsen and Wallevik who produced Olly Murs' 'Up', the British duo Blonde and Calvin Harris. A further entry (Olly Murs' 'Wrapped Up') was the product of a studio owned by established songwriter/producer (Steve Robson's Northern Sky) based around Pro Tools and an API box, a small-format recording/mixing console specifically designed to provide features such as pre-amps, input signal processing, cue sends for high-end (but small) DAW led project studios.

The chart also points to the fact that DAWs have led to a more dispersed production practice in which the sites of creativity are fluid in terms of time, location and the types of technology being used. 'Don't' by Ed Sheeran, one of the few traditional singer-songwriters included in the chart, was originally recorded in multi-platinum selling songwriter/producer Benny Blanco's Pro Tools-based studio in his New York apartment before having additional production from Rick Rubin in Malibu, California. Similarly, although the majority of Sam Smith's three entries on the chart were recorded in one of the UK's leading independent studios RAK, they include elements recorded in songwriter/producer Jimmy Napes' writing studio and further electronic drum production from the hip-hop duo Mojam. Likewise, the Anglo-Irish boyband One Direction's two entries on the chart were almost entirely recorded in hotel rooms on the producer's Logic set up on a laptop as the band fulfilled touring and promotional commitments,[5] an increasingly common practice as music companies attempt to maximize the potential of pop acts that may have a relatively short commercial lifespan.

Thus, this snapshot of contemporary pop production is revealing in that it highlights a number of differing contemporary working practices common within the recording industry. It illustrates the integration of DAWs into the professional practice of the recording industry which has manifested itself in three main ways. First, there are a number of tightly focused pop production teams who have become trusted hit-makers within the global industry. Secondly, there is a more dispersed recording practice whereby different aspects of the production process take place across differing locations, in differing types of production environments. Thirdly, there are a number of solitary producers working in project studios who span a wide range of electronic music genres from mainstream pop and dance to more experimental and avant-garde musical styles.

The pop production teams that dominate the sample chart (in common with many songwriting teams in the contemporary industry) tend to cover key aspects of the creative process from beat making, arrangement and topline melodies that are all undertaken within the DAW environment. Such teams tend to work out of their own studios. Crucially, even at the highest level, the studios tend to differ from the traditional large studio that had been common within commercial studio sector. Due to the nature of contemporary production techniques they often lack expensively designed or large live rooms. Instead, they tend to use smaller spaces for recording vocals. For example, in an interview about equipment, Mich 'Cutfather' Hansen, who

---

[5]See, for example, the footage of the band recording in the NBC TV special ('One Direction: The TV Special' broadcast 23 December 2014).

has scored numerous hits with artists such as Kylie Minogue, Pussycat Dolls and Blue, commented:

> Hansen: Mainly we work on Logic – and then all we need is a good vocal booth and a good vocal mic. As long as the rooms have a good sound, studios can be very basic today.
> Interviewer: What equipment can't you live without?
> Hansen: The Sony mic C-800G and the Mac computer.
> Interviewer: So you don't have any preferences for old school analogue equipment?
> Hansen: I think the type of music we do can all be done on a laptop. (Bouwman 2009)

The way in which modern pop production intersects with the global industry is telling in terms of the centrality of DAW technologies. In this mode of creativity there is clear elision between production and songwriting as many of the building blocks of a particular track are put in place while the songwriting is being undertaken. For example, when working with collaborative partners, Benny Blanco works either in the project studio located in his New York apartment or on his laptop computer even if he has access to a high-end professional studio during a project. He comments on the portability of personal computers and software thus: 'I don't like big studios, and … we usually have a Pro Tools setup in the lounge [of the building]' (Tingen 2012). Describing his own practice, he comments:

> Sometimes I'll start with a keyboard, sometimes I'll start with drums, which I program by dragging and dropping into Pro Tools. I've built up my own sound library over the years. ... I rarely use samples from other songs, unless I need something I absolutely cannot recreate and the sample embodies exactly what I need. But normally, I simply treat my own sounds and my own playing as if it is a sample. (Tingen 2012)

Crucially, this type of approach is indicative of how the trial and error experimentation of the songwriting process often takes place inside, and is in many ways facilitated by the functionality of, the DAW environment. Within this mode of production there is no clearly defined demarcation between differing elements of creative practice. Often demos are worked up in project studios, either with collaborative songwriting partners or before being touted to particular artists or labels. Crucially, the interoperability of DAW environments allows for a layering of sound between specific locations. The starting materials that are produced during these initial writing sessions will often appear on finished recordings.

While this type of layering is central to the modus operandi of a particular production team, in others, the interoperability of recording platforms gives producers a choice of sonic materials recorded in a variety of differing environments. For example, Sam Smith's 'Stay With Me' was a UK number one and US number two single in 2014 produced by Steve Fitzmaurice and Jimmy Napes. The final record is a composite of performances and programming that took place in Napes' composition studio and RAK, a high-end and very well-known London recording studio. Fitzmaurice explains:

> The original session was a songwriting demo recorded at Jimmy's studio. ... [I] tidied it up, and I also added a number of drum samples, particularly bass drum samples [in Napes' studio]. ... I had also made stems of my mix of the RAK version. The final version of 'Stay With Me' ... is a hybrid of the stems from the demo version and a few stems from the RAK version, plus some samples that I added. But about 90 per cent of it is the demo version. (Tingen 2014, 44)

One of the record's defining sonic characteristics was a product of inaccuracies in editing when demoing the song. A drum loop was quickly constructed in Napes' studio which was slightly too long and included a slightly drifting beat in terms of timing:

> The drums on the demo were played by Jimmy and recorded with just one overhead microphone, and he had quickly made a loop of a section of the drum recording he liked. The loop wasn't quite right, it wasn't tied to a grid, and so the timing was slightly off. The other instruments reacted to this, and altogether this created something magical that we couldn't get in any other way. (Tingen 2014, 44)

While the recording of an entire drum kit with one overhead mic clearly subverts traditional mainstream recording practice, the interoperability between studios allows the drum track to be worked up to a releasable standard without losing the 'feel' of the original.

Flexibility

The ability to compose, record and mix recordings across a variety of different sites points to an increasing flexibility in the production process which permeates a significant sector of pop production. The DAW allows for a hitherto impossible level of interoperability in which the materials that go towards final released recordings are worked, reworked, edited and finessed efficiently over a number of complimentary sites of production. In addition, DAW technologies have facilitated material shifts in the working practices of mainstream pop production. Digital technologies are central to this type of freelance project-based working in ways that allow songwriter/producers to capture ideas quickly and efficiently, easily reuse material, work on multiple projects simultaneously, easily compartmentalize between projects, work flexibly according to their own schedules and the schedules of major pop performers and to make pitches to labels and artists through songwriting and production projects that can be worked up and changed expediently in response to collaboration and feedback.

The level of dispersal in terms of creative activity suggested by the sample in the chart is thus indicative of changes in how we might conceptualize the idea of a recording studio itself. It suggests that the increasing fragmentation of recording practice means that no longer can we think of the recording studio as a singular geographic location where the major facets of production are carried out in one place. Rather, differing types of location are used on a task-to-task basis depending on budget, schedule and the specific working practices of an individual producer. In turn, the widespread adoption of the DAW has had significant implications for the shape and size of the commercial studio sector. As Leyshon (2014) has pointed out, the viability of the traditional studio model has become increasingly less financially sustainable since the widespread adoption of the DAW. His analysis of the sector showed that rental rates for studios had not significantly increased from mid-1980s to 2006 and further, that 'day rates' advertised by studios rarely reflect the actual fees paid by clients. Kirby's (2016) research into the UK recording sector bears this out, suggesting a steep decline in the number of commercial recording studios in the UK and a move towards the clustering of smaller DAW-based project studios in managed workspaces as a significant aspect of professional practice.

The clear decline in large commercial studios is also related to the decline in recording industry revenue in the wake of the digitization of sound carriers. In this sense, the shifting economies of scale within the recording industry and shifting modes of production are imbricated and interconnected. In the same historical period as the rise of the DAW, income from the sale of recorded music has decreased and investment in recording has become even more concentrated and has declined in real terms. For example, the IFPI analysis of trends in the first decade of the millennium gives a negative assessment of the effect of digitization upon musical practice, suggesting that there was a 17 per cent fall in the numbers of people employed as musicians in the United States during the ten-year period between 1999 and 2009 and that there was a 31 per cent decline in the value of the global recorded music industry overall between 2004 and 2010 (IFPI 2011, 5). This sharp decline resulted in a restructuring of common financial practice within the recording industry. Perry (2010) estimates that in the late 1990s the average recording budget for a major label project ranged from $75,000 to more than $500,000. By the late 2000s this had significantly decreased to $30,000–100,000 'with only a few established or superstar artists being allocated recording budgets that extend well beyond the $100,000 range'. The move away from large studios towards interoperability, producer-owned studios, trusted production/songwriter teams and in the box production is therefore hardly surprising in this context. The creative strategies and working practices facilitated by

DAWs enabled artists and producers to respond to the new financial realities of the recording industry successfully and flexibly. Interoperability allows for expensive time in high-end studios to be kept to a minimum, the layering of sonic materials within the DAW environment allows for investment in polishing or finishing a product as the project requires and songwriting partnerships and artist/producer relationships can be set up quickly and efficiently in immediate response to success in the market.

In many ways the production practices outlined here can be understood as a form of flexible specialization. Piore and Sabel (1984) initially coined the concept of flexible specialization to refer to developments in the 1970s whereby large corporations moved on from a mass production model to employing technologically astute artisanal methods through small units that could react to changes in the market, target niche markets and be financed or discontinued accordingly. In relation to cultural production, the term has been used to account for how large corporations have used such strategies to maintain a 'creative vitality' (Lampel et al. 2000, 263) that is crucial in creating the necessary turnover of new cultural products and also as a means to manage unpredictability in terms of audience demand. Strategies, including 'outsourcing' to independent music producers, enabling the major labels to appropriate niche genres by drawing on their specialist expertise (Tschmuck 2012, 263) and using relationships to smaller companies as a form of market research (Bader and Scharenberg 2009, 88), have been read as indicative of the recording industry's move towards flexible specialization from the 1970s onwards.

However, as Hesmondhalgh's (1996) critique of the field notes, flexible specialization accounts have tended to be overly deterministic to wider industrial patterns, giving a historically skewed picture of developments within the recording industry in the 1970s and 1980s. As he points out, there has been a 'longstanding importance of the relationship between specialized niche markets and the mass pop market' (1996, 483). For example, Peterson (1997, 187–92) notes that the success of genres such as country music and rhythm and blues from the 1920s to 1940s signified somewhat diffuse patterns of creativity within American recording industry whereby songwriting and production often took place outside of the institutionalized control of Tin Pan Alley and the major labels. In fact, even in the song factories of the Tin Pan Alley system, creativity was managed in the form of tightly knit small-scale songwriting teams that could be funded or abandoned according to their relative success. Significant aspects of the music industries then (especially in relationship to creativity) have to some extent *always* been highly flexible in terms of the way they intersect with the structures of large entertainment corporations. We should therefore see the predominance of songwriting/production teams in the current market as keeping with patterns of the bureaucratic management of artisanal creativity that have been characteristic of the recording industry since its inception.

The picture of mainstream pop production outlined here is clearly more complex than the democratized, decentralized environment that is often claimed for the contemporary post-digital context. It is largely dominated by a fairly small group of trusted creative workers who are situated professionally, financially and socially within the structures of large entertainment corporations. Such production teams and producer/songwriters have the necessary industry contacts and levels of cultural and professional capital that enable them to operate at the highest level of the recording industry. Their professional relationships are enmeshed within the strata of cultural intermediaries within the recording industry such as A&R, business affairs and marketing staff who 'come in-between creative artists and consumers' who are 'continually engaged in forming a point of connection or articulation between production and consumption' (Negus 2002, 503). While being reflexive and responsive to perceived trends and changes among consumers, such cultural intermediaries nonetheless fulfil roles that serve to maintain boundaries around the professional field.

The position of these high-level production teams can be understood as semi-autonomous. They own and maintain their own studio spaces and have a high degree of control in terms of the minutiae of creative practice. They also are not contractually bound exclusively to one specific record company, instead working between companies from contract to contract. Nevertheless, they have to maintain contact with cultural intermediaries on a variety of levels throughout the production process. Their modus operandi is dependent on these flexible yet highly professionalized relationships. Hence, despite the actual technologies and contexts of production shifting because of the rise of the digital audio workstation, we should understand these practices as being a continuation of existent and normative set of organizational relationships that have existed within the recording industry through a large proportion of its history.

Further, the type of practice indicated by this indicative example of contemporary production contexts constitutes only a partial decentralization of practice. In the majority of cases despite utilizing digital technology where recording can be made anywhere, these creative workers tend to be located in existing centres or hubs of the global industry. For example, thirteen of the entries in the chart sample were produced in London and nine in Los Angeles. While the proliferation of Scandinavian producers within the top level of international co-production can be read as a move away from the traditional recording centres of Los Angeles, New York and London, their track record over a significant period of time has meant that Copenhagen and Stockholm have now become integrated into the international landscape of the upper echelons of the recording industry.[6] The recording locations in

[6]Even then, it is perhaps telling that Scandinavian producers such as Avicii and Max Martin have relocated to Los Angeles.

the sample indicate that it is in the dance music sector that we can see the most geographical dispersal away from the centres of the recording industry (provincial UK cities such as Nottingham and Bristol along with various European and US locations). However, this is in keeping with the history of electronic music since the 1980s where project studios were located in a variety of geographical locations such as the major cities of the north of England, Belgium, the Netherlands and Italy (see, for example, Straw 1991; Hesmondhalgh 1998). Again, we can see the production of electronic dance music in the post-digital era as congruent with patterns of flexibility and incorporation that have historically been a core characteristic of the recording industry. Here, local 'agglomerations of flexible production' are appropriated and integrated by transnational corporations by taking over 'those segments of music production that benefit from concentration, i.e. [promotion], reproduction, global distribution and global brand name' (Bader and Scharenberg 2009, 85).

Overall, the analysis presented here is indicative of how big media has appropriated and adapted to the changes in production ushered in by digitization in ways which are expedient to their maintenance of existing power relationships. The (surviving) major labels have, in effect, learnt to swim in the new online seas through a digitalization of their business practice. The analysis undertaken here illustrates quite graphically that large music companies rely on producer/songwriters to come up with the musical goods for stars. In other words, the star-making machinery that has been a constant within the recording industry still exists and still counts. However, the cost structure of producing popular music hits has changed; the majors no longer control access to (expensive) recording and (expensive) distribution and their costs are now whatever royalties and flat fees they pay producers.

Further, digital production technologies have allowed a more flexible set of production practices and a new (and relatively select) breed of producer/songwriter to become the norm at the commercial top end of the industry. Because of interoperability, collusions around recording tracks become increasingly more viable. In turn, the more efficient the colluders, the more successful these producers (and musicians become), the more that these practices become the norm. In this way, digital production technologies have allowed music companies to respond to shifts in the music economy more widely that have led to a shrunken market for sound recordings in which there is a user-preference for tracks over albums, and EDM over rock. As I suggested in the introduction, the digitalization of the recording industry has entailed a move away from long-term investment in core genres towards a concentration on cash-cow styles and a series of strategies through which it can instantaneously respond to current trends. The importance of these differing factors is illustrative of how the multiple effects of digitization converge to produce social realities. Changes in production technology thus have to be understood in the context of wider experiences of political-economic change, but it is also important to consider how these specific technologies have served to facilitate and focus that change.

Summary

This chapter has argued that rather than providing a wholesale democratization of music practice, digitization has had more diffuse and complex effect upon differing areas of music production and music industries. While it is clear that the advent of computer-based DAW technologies and Web 2.0 has meant that music technologies are more accessible, more individuals are involved in music production and there are numerous ways to connect and disseminate this work, the structural organization of the music industries has developed in ways which we cannot simply read out as democratization. Rather, the structural relationships of popular music, both in terms of large corporations and the independent sector, have adapted to, and accommodated, new technologies in ways that to a large extent maintain their economic and cultural dominance.

This is not to say that the effects of digitization are not profound, rather that they should not be unproblematically celebrated as democratizing within themselves. However, this chapter has outlined specific areas such as the democratization of knowledge and the wider adoption of studio technologies that have significant real and potential effects popular music practice. For example, the recent emergence of a number of successful young female producer/musicians who began their careers through experimenting on garageband (Tavana 2015) perhaps hints at the potential of digital technologies in eroding, or at least bypassing, the traditionally 'masculinist' and exclusionary space of the traditional recording studio (Leonard 2007). Similarly, dubstep and grime (as discussed in Chapter 5) are just two of a number of important forms of cultural expression directly facilitated by the availability of cracked software and the inherent usability designed into the functionality of DAWs. These genres are indicative of how digital technologies have the potential to provide a platform for otherwise marginal voices. Stobart's (2010) ethnomusicological work in Bolivia suggests that digital recording and reproduction technologies have enabled the emergence of music produced on a small scale targeted at an emergent low-income indigenous market, allowing for a plurality of cultural voices which 'stress a diversity of regional, local, urban, rural or ethnic identities' and 'offer contrasting music aesthetics and values; and they often reflect internal national issues and struggles' (2010, 48–9). It is at the small-scale levels of production that digitized music technologies have the potential to create a discursive cultural space that is genuinely progressive and engage with wider power relationships. Having said this, there are still clear issues of access in this regard. Despite the emergence of numerous high-profile DAW-facilitated women producers for example, cultures of digital music production are nonetheless male-dominated and subject to masculinist exclusionary discourses (Bell 2015).

What is clear from patterns across all levels of popular music production, from amateur bootleggers and studio hobbyists to the highest echelons of the recording industry, is that the DAW has had a profound effect upon creative practice. It has shifted what it means to be a producer, writer or artist and it has fundamentally changed how musical creativity is undertaken for many producers. It is this shift in the minutiae of day-to-day creativity and how it is experienced that forms the core of the next two chapters. The next chapter explores the emergence of the computer-based DAW and how its historical entwinement within concurrent developments in the personal computer has led to new creative modalities within the production of popular music. Chapter 3 expands on this historical trajectory by exploring how the constructed functionality of the DAW leads to specific ways of working and a significant shift in the conceptualization of music in its creative context.

CHAPTER TWO

Affordance, digital audio workstations and musical creativity

Computer Chronicles, *San Mateo California, October 1986*

In a 1986 episode of the US Public Broadcasting Service (PBS) network television programme the Computer Chronicles *Bob Moore of the software developer Hybrid Sounds Inc. is demonstrating his company's audio application ADAP (an early computer-based recording and sampling system) to the US computer scientist Gary Kildall. As Moore demonstrates how the system can capture audio from an external source the two have the following exchange:*

> *BM The recording system takes the analogue information from here [points at CD player] digitizes it, saves it as RAM and then I can play that back and manipulate that information. … I can grab a section if I want to, I can zoom in if I want to.*
>
> *GK So this is sort of like the equivalent of a word processing system but for music instead?*

BM Right! I can grab a section and place it anywhere else in the song that I'd like to.[1]

Bleu Remix, Glasgow, National Review of Live Art, 2008

I am watching a man turn blue. To be more accurate, I am soundtracking a man turning blue. I am in Glasgow for the National Review of Live Art to perform with the Swiss performance artist Yann Marussich. Marussich sits in a specially constructed light filled glass tank dressed only in a loincloth as part of a project called Bleu Remix, a number of collaborative performances with several European musicians for which Marussich would win the award of Distinction Prix Ars Electronica: Hybrid Art later in the year. Over the next hour a blue liquid will progressively seep out of his eyes, mouth and nose before eventually emerging from the pores in his skin until he becomes entirely blue. During the whole performance Marussich will remain entirely still.

Two weeks earlier Marussich had sent me a DVD containing over a gigabyte's worth of files. These were audio recordings of various parts of the artist's body captured by contact and stethoscope microphones. As the files are labelled by number rather than source, I organize them according to their sonic qualities: rhythm, texture, amplitude etc. and begin to experiment with how they might fit together. Despite my initial groundwork, my own performance in Glasgow is very much an improvisation, an attempt to respond to and frame the artist's feat of physical endurance in the moment. During the hour-long duration of the performance I am in a state of intense concentration. I am visually focused on Marussich's body but am simultaneously using hearing, touch and sight to alter the sound in the room. I drag and drop files into differing audio channels on the screen of my laptop, trying to react to the nuances of his body's reaction

[1]Computer Chronicles MIDI Music. Broadcast 01/10/86 available from the Prelinger Archive. http:www.archive.org.

to the dye. I use filters to gradually fade in elements of the soundscape, loop elements of the audio files into rhythmic patterns and use differing audio effects to alter their textures. My movements are gradual and subtle as I try and lock into the slowly evolving arc of the artist's physical transformation. I progressively move from miniscule droplets of sound through to an engulfing blanket of muffled sub-bass heartbeats and noise as the hour progresses. As Marussich's skin becomes streaked with perspiration, eventually completely glistening in the blue of the dye, I gradually change frequencies using assigned knobs and faders on my MIDI control surface. As the process of transformation comes to an end I build up the bass frequencies to create an inescapable imposition of the sounds of Marussich's body upon the audience. As the bass frequencies increase in volume becoming almost unbearable, Marussich's stage manager gives me a previously agreed signal. I stop the audio as the performance space abruptly descends into complete darkness.

On the surface, these two short epigraphs might seem initially at odds. One, a rather prosaic throwaway observation on an early version of a type of computer program that would become ubiquitous over the next two decades is perhaps typical of the banalities of low-budget television chat. The other, a vivid creative experience of my own, is far from banal. I am able to recall the details of my experience with such clarity precisely because of its exceptionality. Yet the two are intimately connected. The latter could not have happened without the former and the intensity of my experience was facilitated by my use of, what for me has become as much as an everyday technology as a word processor or a spreadsheet program. In fact, Kildall's analogy of the word processor is at once an easy shorthand to explain the functionality of the program he was describing to his viewers and a succinct synecdoche for the way in which musical creativity has become integrated within wider modalities of personal computing over the past thirty years.

On a personal level, performing the audio elements of Bleu Remix that night *was* an intense creative encounter that felt intuitive as an embodied experience. It involved sight, touch, hearing and bodily vibration caught up in such a mental concentration that I felt physically drained afterwards. Yet my improvised soundtrack was facilitated by the sound technologies I had at my disposal. I was using Ableton Live, one of the most popular pieces of musical software on the market, to order, process and manipulate

Marussich's bodily sounds. The technology itself was clearly an actor in the creative process and my artistic response was undoubtedly shaped by its capabilities and limitations. The program served to organize my thinking about the creative outcome of my endeavours in very particular ways. Its particular set of functions and audio effects offered a clear set of sonic possibilities from which I made a set of decisions about what to do with the raw audio material. Its spatial organization perhaps had a subconscious effect upon the ways in which I understood the particular order of the sounds, and its graphic interface shaped my conceptualization of how parameters of frequency and effect were altered. Yet at the time, my engagement with the creative task at hand felt instinctive, almost naturalized, with the technology itself rendered transparent. In short, the technology both shaped my creative trajectory within the project while allowing for a freedom and spontaneity that *felt* natural.

It is this seemingly contradictory nature of digital technology that provides the focus of the following chapters. The chapters examine the most significant development in digital music making since the 1980s, the computer-based digital audio workstation (DAW)[2] in terms of pragmatism, limitation and creative freedom, as an everyday technology that has the capacity to be spectacular and transformative. The chapter in hand suggests that this very specific musical technology has to be understood within the wider context of our everyday uses of technology in an era of increasingly ubiquitous computing. It therefore traces a historical narrative which links the emergence of a particular design logic within computer-based music software to wider trends in the aesthetics and functionality of computing and their combined effect upon new creative modalities for musical practitioners.

Mainstream digital audio workstations

It should be noted that the technologies and processes under examination within this chapter (and more broadly in this book as a whole) are those that are most widely used across popular music production and are at the centre of a multitude of differing genres and practices. These 'mainstream' computer-based DAWs such as Pro Tools, Cubase, Sonar, Logic, Ableton Live and FL Studio have their roots in the recording, synthesis and sequencing hardware that had become central to the creative process within the popular music industry immediately prior to the exponential growth of the personal computer industry in the 1980s and 1990s. As a result, they tend to replicate the key tasks of this hardware technology (synthesis, recording, sampling

[2]The concentration here is on computer-based DAWs as opposed to integrated DAWs: that is, hardware DAWs which integrate audio processing and conversion, a control surface and data storage into a singular device.

and sequencing), although, as we shall see, their situation within a broader computer culture and the integration of these tasks into singular integrated software applications means that they take on conventions of their own. They also share a common representative approach to their individual tasks and have similar spatial, graphical and functional dimensions. As such, they fundamentally differ from the software used within the algorithmic and generative approaches to composition and performance that have been dominant within institutionally supported academic electronic music, avant-garde and experimental traditions.[3] The uses of technology within the academic electronic music tradition have a different historical trajectory and are resultant in a different set of modalities as the programs used tend to be 'procedural rather than declarative' and thus have a different set of 'abstraction mechanisms' (Duignan et al. 2010, 22) (i.e. ways of representing sonic and musical processes). While the conventions and practices of this tradition have been written about comprehensively elsewhere (see, for instance, the approaches covered in Collins and d'Escriván 2009 or much of the work contained in the journal *Organised Sound*), such work has invariably been from inside the institutional context of post-electro-acoustic academic composition and is as such heavily informed by its discourses and praxis. Such work is often active within the construction of a hierarchy that ignores the technology central to the mainstream of computer-based musical production or relegates it to a secondary position. As a result, there is something of a conflation within scholarly writing on the processes of 'computer music' and the development of technologies relevant within the fairly narrow confines of the academic/avant-garde tradition.

This is something of an oversight as these mainstream DAWs have had a cultural impact that has spanned musical and audio production from the macro to the micro. They are at the heart of high-end professional studios that produce multimillion-selling records and the same time they are central to a more diffuse proliferation of amateur and underground cultural production and are central in the rise of the audio prosumer. They are widely used in the film industry to produce soundtracks and Foley within productions with enormous budgets and simultaneously in the zero-budget remix films that regularly receive millions of views on video sharing sites. They are the focus of a multimillion-dollar computer software and hardware industry, and are illegally traded as cracked software by teenagers throughout the world. In short, they have left a stamp on a broad range of practices that has left the world of musical production irrevocably transformed. They have changed the structure of the entertainment industries and, in a large number of instances, the terms of engagement between the creative individual and sound.

[3]It should be noted that DAWs such as Logic and Pro Tools are also widely utilized in contemporary electro-acoustic music and that Ableton's collaboration with Cycling 74 on the launch of Max for Live has served to integrate this approach within mainstream DAWs.

At the heart of what follows then is an interest in how creativity within digital music cultures takes place within a human technological nexus of interaction. While Chapters 3 and 4 examine the creative process from the perspective of musicians and producers in more depth, the main focus of this chapter is upon the technology itself. It suggests that there are properties intrinsic to DAWs in terms of their functionality and interfaces that have a fundamental bearing on what musicians and producers actually *do*. In other words, both human and technical actors (Latour 1999) have agency within the creative process. Secondly, the chapter attempts to lay a theoretical framework suggesting that this network is also crucial in understanding what creative individuals *feel* during the creative act. Much of the chapter is therefore given over to the discussion of how the design of the graphical user interfaces (GUIs) of DAWs has developed; as Bolter and Gromala (2006, 22) note 'to design a digital artifact is to design an experience'.

In order to unpack the modalities of music making engendered by computer-based production, it is important therefore to consider the experiential properties of the DAW as a device. In particular, the chapter will concentrate upon the (constructed) idea of intuitiveness, interface design and the sensory organization of creativity produced by such applications. Clearly, my overall experience that night in Glasgow was one of intuitiveness, as Chapter 4 suggests, a not uncommon feeling for those engaged in particular creative encounters and moments. It has become a truism of the music technology world that the design of the interfaces of DAWs tends to be much more 'intuitive' than in hardware-based production technology. The idea being that DAWs create a technological transparency whereby there is less mediation between technological tasks used in order to realize a creative idea and the actual idea. In this sense, DAWs are a part of much more pervasive thrust in the computing industry that has framed every new technological development from the PC to the iPad. They are intimately situated within wider techno-social networks and commonalities of human–computer interaction (i.e. within systems that 'enable their users to interact with digital information' Visell 2009, 38) that have emerged through the ubiquity of personal computers. Sketching a more general framework of computer interaction, therefore, is important in understanding the trajectories of creativity in a culture infused with digital technologies and a musical environment where computer-based production has become the norm. In order to unpack this a little more, I would like to think about the GUIs of DAWs in terms of intuitiveness, affordance and transparency.

Affordance

Emerging within the field of social psychology, the concept of affordance has been influential across a number of subject areas relevant to this book, including human–computer interaction (Norman 1988; Turner 2008), music (Clarke 2005; DeNora 2005[4]), and music education (Pea 1993; Gall and Breeze 2005) and it is a concept which will recur in subsequent chapters. Affordance is used as a central theoretical concept throughout the book as it allows for a holistic approach to creativity in this context. It simultaneously allows for an understanding of musical and sonic materials, the technological tools used in composition, and the intersection between the two. The relation between human, action and environment is thus crucial in trying to come to terms with new modalities and ways of thinking relating to the use of virtual studio technology (VST) and DAWs. The conceptual core of what follows is that creativity within this new digital context of musical production is underpinned by differing types of affordance. This chapter argues that these technologies offer certain affordances that are related to common modalities within HCI (Norman 1988, 1999). Chapter 3 goes on to argue that the sounding object has certain actionable properties (Clarke 2005; Kreuger 2010) that are central to creative action for electronic producer/musicians.

The term 'affordance' was initially coined by the perceptual psychologist James Gibson (1979) as a way of describing the actionable properties of a given object as perceived by human or animal actors. Within this theorization an individual's perception of the environment is resultant in particular courses of action whereby affordances indicate the range of possible activities of a given object. If we take the simple example of a paper cup, we can see that objects have affordance in a number of differing ways. Its shape clearly demarcates it as a vessel suitable for holding liquid. Its ergonomics and circumference suggest a graspability which is attuned to a human hand, its paper holder suggest where that hand can grasp without fear of burning and its paperness suggests its disposability. In an instant we can make quick assessments about the object in terms of functionality, mode of use (the canonic affordance or ordinary orientation with which we interact with a given object, see Bærentsen and Trettvik 2002; Borghi and Riggio 2009), temporal convenience and even the appropriate occasion in which to use the object.

Affordance and software design

Although conceptual framework of affordance emerged within the field of perceptual psychology, its major significance to this chapter is through its practical and discursive application. A tacit understanding of affordance

[4]These accounts from musicology tend to use affordance as a tool to conceptualize music as a material which is active in structuring embodied or cognitive experience. While these accounts provide a valuable way into thinking about the affective and bodily power of music, my main concern here is with how creativity is experienced. As such, accounts from HCI studies provide a more pertinent framework for the discussion within this chapter. However, the idea of music as an affordance is picked up in more detail in Chapter 3.

has been the key driving force within the field of design, and ever since Gibson made his initial theoretical formulation, the concept has been widely appropriated in the fields of ergonomics, product design and computer programming. For example, Norman's (1988) conceptualization of affordance as the design aspect of an object that offers an immediate visual indication of its functionality became a widely quoted and followed dictum underpinning the logic of computer interface design:

> Affordances provide strong clues to the operations of things. Plates are for pushing. Knobs are for turning. Slots are for inserting things into. Balls are for throwing or bouncing. When affordances are taken advantage of, the user knows what to do just by looking: no picture, label, or instruction needed. (Norman 1988, 9)

This appropriation has formed part of a discourse of usability that has been central to the computing and software industry since it began to realize the commercial and cultural potential of information technology. The fact that the term 'user friendly' and the concept of 'information architecture' emerged as guiding principles for computer software developers in the early 1970s (Mulvey 1979; Ding and Lin 2010) is indicative of how the industry was seeking to cross over to a business and leisure market at the time. The advent of the microprocessor had opened up the possibility for more compact and widely affordable computing systems, and as a result, programs had, as a necessity, to develop a more interactive form of computer use that would be accessible to the non-specialist (Press 1993). In short, it was in the commercial interests of the computing industry for interfaces to be understandable to as wide a caucus as possible. Bolter and Gromala (2006, 376–7), for instance, note that the Apple corporation began to 'commodify' the idea of 'transparency' in application design from an early stage and that such discourses were apparent in the company's interface guidelines as early as the 1980s. The link between design aspect and the appearance of immediate functionality then emerges in this very specific techno-commercial nexus, and over time, the holy grail of interface design became the production of a virtual space with clear affordances, which meant that applications could be intuitively explored by the non-specialist (Turner 2008).

Intuitiveness and DAWs

The discourse of affordance found its way into received logic (and everyday parlance) within the computer industry through the linked concept of intuitiveness. Turner (2008, 475) observes that intuitiveness is perhaps 'the most fundamental, most desirable of attributes of human–computer interaction' and yet the computer industry has been 'remarkably vague on agreeing what is meant' by the term. Nevertheless, the term has become such a key marker within computing culture that it merits a separate entry in the *Oxford English Dictionary* as: '(chiefly of computer software) easy to use and understand' (*OED* 2006, 909). This rather flat, prosaic definition stands in contrast to the more widely applied meanings relating to knowledge and mental perception. The idea of acting, perceiving or reacting 'without conscious reasoning' or 'in immediate apprehension, without the intervention of any reasoning process' (ibid.) strikes me as fundamentally different from merely 'easy' or 'understandable'. However, there is an inevitable erasure of this wider meaning when intuitiveness is applied to HCI. Despite the rigidity of the *OED* definition, in actuality the usage of the term oscillates between the broader and specific definition and is clearly employed to *imply* the broader sense of the word, especially in terms of software developed for creative purposes.

In line with their development within a wider design logic of HCI, intuitiveness has thus become a key selling point within the DAW market. A survey of the way in which the major software companies describe their applications bears this out. Ableton's blurb aimed at educators for instance, begins with the statement: 'With its intuitive, powerful design, Ableton Live opens the world of music creation to every student, regardless of age and ability in a way that's straightforward and a lot of fun.' Likewise, Steinberg's website boasts that the company 'is taking sequencer-based composing to a whole new level and offers the most intuitive and functionally complete MIDI composing and sequencing toolset on the planet'.[5] Although ease of use is clearly a selling point here, a second way in which the intuitive nature of an application is marketed is in terms of creativity. Intuitiveness as positioned in the discourse of DAW developers is based in selling the idea of a kind of creative disintermediation[6] whereby the intuitiveness of the application means there is less distance between a musical idea and the realization of that idea. For example, Logic Studio is explicit in relating the purported ease of use of the program in terms of the user's creativity. Apple's introductory webpages on Logic Studio are littered with statements such as: 'Whether you're improvising a three-chord pop song, making beats, assembling and reshaping loops, or composing for video, nothing will break your creative flow' and 'it's easy to get amazing sounds and amazing-sounding recordings. Now you can tackle any stage of your project yourself – without losing your inspiration along the way.'[7]

---

[5]http://www.steinberg.net/en/products/cubase/cubase6_details.html (accessed 16 January 2011).

[6]This idea of creative disintermediation has been a part of the discourse of the computer music industry throughout its history. For example, a 1992 edition of the US PBS television program the *Computer Chronicles* contains a short piece about computer music research (CCRMA) at Stanford University. While explaining the radio baton, a MIDI controller he is developing, Max Mathews explains that 'because the technique is in some sense easy on the instrument it allows the performer to focus his [*sic*] attention and mental abilities on the expression or on what he wants to say with the music. What he wants to say can be called the soul of the music.' http://www.youtube.com/watch?v=ffy2Bo82N9s&NR=1 (accessed 17 January 2011).

[7]http://www.apple.com/logicstudio/what-is/ (accessed 15 January 2011).

These examples are symptomatic of a general thrust in the discourse of computer technologies in that the more advanced a technology becomes, the more transparent it is purported to be. For example, within Scandinavian activity theory which emerged in the 1990s, an application is seen to become transparent 'when the user is able to direct conscious actions to the object of work (e.g. the novel the writer is working on) whereas the computer application (the tool) is handled through non-conscious operations' (Bertelsen 2006, 360). Yet there is something more, in that these are software applications aimed towards creative practitioners. The focus here is upon individuality, creativity and interpretation rather than technique. DAWs provide tools, not to be mastered in terms of virtuosity or expertise, but rather to provide a simple and direct facilitation of an individual's creativity. The use of intuitiveness here serves to appeal to the creative identity of the consumer by promising to provide the context of creative and intuitive action, to encourage 'inspiration' and 'flow'.

Such an appeal is thus underpinned by how the idea of intuitiveness is also is embedded within everyday discourses about what creativity actually is. Ongoing, pervasive romantic notions of creativity continue to perpetuate the view that 'creativity has a lot to do with the extraordinary and the use of innate gifts of intuitive talent' (McKintyre 2008, 40). In addition, intuitiveness also relates to how creativity is experienced by the individual involved in a particular artistic practice. As various studies (Csikszentmihalyi 1975; Negus and Pickering 2004; Leonard 2010) have shown (and as is explored in more detail in Chapter 4) many artists working across a variety of media often describe the creative experience as 'being so caught up in the making of their art that they seem to be taken over, as if ... [the creative act is occurring of its] own accord' (Negus and Pickering 2004, 19). Both, everyday discourse about intuitiveness and creativity then, have implications that go beyond rationality towards instinct and feeling, and the two are often connected or implied in software industry discourse. The 'intuitive' application facilitates or unlocks creativity within an individual and the implications of the term go beyond mere ease of use. When applied to 'creative' applications, intuitiveness is thus caught up in a complex matrix of conceptual analogy. It draws upon the specific modes of emotional and compositional engagement that the experience of interacting with new technologies elicits. Despite, and perhaps because of, this rather knotty set of associations, I think that intuitiveness remains a useful way into thinking about how creativity is experienced by electronic musicians and producers.

Familiarity and design conventions

The creative disintermediation implied by such marketing materials is indicative of an underlying rationale that underpins the design of DAWs. They are designed with affordance as an organizing principle in order to be accessible and to be easily explored by their users, so that a wide variety of tasks and functions can be undertaken quickly without the prerequisite of a steep learning curve or *a priori* technical/acoustic knowledge. Gall and Breeze's (2005) study of school children's interaction with music software in the context of a compositional project to some extent bears this out. In a series of empirical exercises, students were not given specific information about how to use the software by teachers but were nevertheless able to competently handle its features from an early stage. The study noted that the particular software the students were using (Ejay and Cubasis) was built around 'contemporary signage' (2005, 421), that is, design aspects that offered clear affordances for action that the children could easily engage with. These findings undoubtedly demonstrate a graspability that can be utilized even by a child 'just by looking: no picture, label, or instruction needed' (Norman 1988, 9). Yet Gail and Breeze's identification of 'contemporary signage' hints at something less immediate, more the product of accumulated knowledge that perhaps gets us closer to unpacking the idea of intuitiveness.

If we return to my original 'paper cup' as an example of affordance, the sheer amount of information contained within computer–human interfaces places a higher level of demand upon their users than a vessel merely designed to carry liquid, and we cannot see the two as constituting the same type of self-apparent functionality. The 'affordances' of GUIs are rather constructed in an ongoing historical trajectory which affects how the visual/tactile conventions of computer interfaces are perceived as intuitive through a process of enculturation. As a result, many critics have made a distinction between physical (or 'real') affordances and 'perceived' affordances (Pea 1993; Norman, 1999). In his 1999 revisiting of the concept, Norman argues that the affordances offered by computer interaction are actually learnt, and that a 'community of practice' develops over time relating to functionality. In addition, because of their location within a human/technical network it is clear that canonical and perceived affordance shift over time and that the development of action in this context is an 'ongoing aspect of the use situation' (Bertelsen 2006).

In turn, the fact that such a constructed version of 'intuitiveness' is so valued by software developers also has clear material implications for creative practice. The modalities engendered by the DAW's particular form of human–computer interaction can be seen to actually impact upon the creativity of musicians and producers in very particular ways, affecting the experience of the creative act and structuring the way in which the creative process unfolds. In order to unpack this fully, it is necessary to place the design of contemporary DAWs within the context of two concomitant historical strands: the gradual integration of MIDI and audio processing into personal computer – based music technologies and the emergence of wider visual/functional conventions that have developed since the rise of the personal computer more generally. Both of these elements are fundamental in understanding how recording, sequencing and compositional technologies have been adapted and reimagined within the context of personal computers.

Screen-based interaction

The most immediately apparent implication of the emergence of the DAW as a key site of musical creativity is that musical information is conceptualized visually. This is part of a wider trend since the 1960s whereby human–computer interaction has become first and foremost screen based. The ubiquity of GUIs which allow computer users to carry out commands using a diagrammatic rather than text-based interface has had a fundamental effect upon how users interact with computers and, importantly, think about interacting with computers. Work within film and new technology studies has noted how differing forms of screen-based media are far from neutral, each engender structured ways of looking. From the concentrated (and often gendered) gaze identified in the darkened cinema (Mulvey 1975; Ellis 1982) to the multiperspecitve interactivity of games (Chesher 2004), the viewer's experience has been seen as being fundamentally shaped by their relationship to the screen.

In turn, the conditions of this experience are further produced by conventions in design that have cohered over the past three decades. From the 1970s GUIs across a variety of divergent task-based applications have tended to demarcate and represent distinct functions through the organizing principle of the window (Blackwell 2006). The idea of the window as an overriding design standard has meant that a certain type of visual thinking has become ingrained within a variety of everyday tasks. Since the introduction of the window-based platforms (such as Mac OS, Microsoft Windows and Linux), there has been a normalization of thinking about a variety of tasks and concepts in terms of windows, from the way we write to the way we do our accounts and how we store and edit the significant (and quotidian) moments of our lives through photography and video.

In turn, the window 'holds' a further set of normative visual metaphors (function-based icons, drag and drop, highlighting, drop-down menus, cut and paste, etc.), which quickly became standard ways of carrying out tasks. The release of Apple's Macintosh in 1984 can be seen as the commercial mainstreaming of the GUI. Significantly, it was a system which adopted the visual metaphor of the desktop with 'folders' and 'windows' that remains dominant to this day.[8] As Freidland (2005, 17) notes, its launch

[8]It should be noted that the desktop metaphor had been in development since the 1970s. See Blackwell (2006) for an even-handed discussion of this development in which he positions the metaphor of the desktop in terms of technical evolution and historical contingency rather than any dominant theoretical position held by researchers.

signalled the movement of the personal computer from being an 'esoteric hobbyist device to a fetishized mass-produced commodity'. The Macintosh was quickly followed by the launch of the Amiga with its very similar Workbench and the Atari ST's GEM (Graphical Environment Manager) (Chinn 1985). The widespread adoption of this metaphorical matrix meant that for users menu bars, pointing, clicking and dragging which would initially have seemed 'strange to new users' (Bolter and Gromala 2003, 46) constituted a new visual orthodoxy: 'So "natural" and "intuitive" to us, that an interface without menus and a pointing device now seemed odd' (ibid.). The early 1980s also constituted a concurrent paradigm shift for the software industry (Cambell-Kelly 2003, 215–18), wherein a series of key productivity applications (spreadsheets such as VisiCalc and SuperCalc and word processors such as WordStar and Word Perfect) served to drive sales within the personal computer market. The emergence of these 'killer apps' (Freidman 2005, 103) proved 'so compelling' that they spurred consumers to buy particular computers on which to run them. The success of these applications was also instrumental in the emergence of a generic visual architecture, an iconicity and functionality that served to normalize particular ways of structuring and manipulating information.

On one level we may see this as having a distancing effect. The metaphor of the window presupposes a separation between the body and subjectivity of a viewer, and a viewed external other: they are interpreted as membranes between 'inside' and 'outside' (Richardson 2010). However, the interactive functionality of the actual uses of windows in the personal computer context serves to complicate such a binary relationship. The semantic split between the film and television *viewer* and the personal computer *user* is significant. Windows also suggests a transparency, an intuitive space that the user can penetrate. The functionality of windows-based applications means that the window presents a navigable space in which the user has agency.

The idea of navigability also leads us to the matter of virtuality; that is, the 'graspablility' in HCI is virtual. The lack of actual contact in the relationship between function and movement means that any physical action that takes place is always essentially a mediated experience. However, as my own experiences described at the start of the chapter testify, they are not felt as such. Rather, they are experienced as an embodied whole whereby the mediating technology almost seems to disappear. This disappearance has long been a design goal within computer interface design. Bolter and Grusin (2000, 31) note 'windows opened on to a world of information made visible and almost tangible to the user, and their goal was to make the surface of these windows, the interface itself, transparent'. Bolter and Grusin go on to problematize the notion of transparency (ibid., see also Bolter and Gromala 2006), arguing that computer interaction can never escape the traces of its mediating technology in a process of 'hypermediacy' whereby the viewer is always aware of the framing devices used to attain the mediated experience. However, it is clear that in our everyday uses of computers transparency and hypermediation do not have to be mutually exclusive. Our engagement with the interactive screen constitutes something akin to the suspension of disbelief in the viewing of screen media such as film and television: yes, we know we are working with a mediated technology but the sense of engagement we feel transcends this self-awareness. The sedimented cognitive acts (Husserl 1973) that characterize our ongoing relationships with computer interfaces mean that most of us will have experienced such deep levels of involvement when engaged with a particular task that there seems to be no disconnection between our thoughts, our actions and technology.

Richardson (2010) identifies a number of body-media relationships that are 'moored by sedimented cultural habits, body-metaphors and tropes surrounding our engagement with screens', in which stance, position and 'the impact of the situated or built environment' have a clear effect upon that engagement. In other words, the basic combination of screen, mouse and track pad differentiates the personal computer from other forms of screen media with regard to engagement. In terms of computer music production, this relationship is further heightened by the integration of control surfaces into the everyday working practices of the producer/musician's interaction with DAWs. However a GUI is controlled via midi interface, it is indicative of a tightening of body/screen encounters. An individuals' interaction with a personal computer is in essence a multisensory encounter in which touch, sight and hearing converge. The combination of these elements constitutes a material engagement that is *experienced* as an embodied whole. Here, the kinetic, visual and aural *feel* connected and simultaneous. Turner (2010) makes a similar point with regard to the way the term 'intuitiveness' has actually been used within computer–human relations. He notes that taken as an organizing principle of design, familiarity, embodiment and action-perception coupling constitute the core elements of intuitiveness in computer–human interaction.

Hence, my feelings of connectedness, intuition and naturalness during the creative encounter described above can be partially explained as being a result of the enculturation of visual/tactile thinking through the organizational strategies of the personal computer. As our interactions with computers have become part of everyday life, our use of tactile interfaces to relate to on-screen visual interfaces has developed into motor skills; that is, that they are being learnt and internalized to such an extent that they are not only becoming smooth and efficient, but are also being effectively naturalized. For example, in their discussion of computer games and ANT, Cypher and Richardson (2006, 5) argue that the conventions of media interfaces within personal computing constitute a sociotechnical network of human and non-human agencies which enforce particular habitual behaviours which are naturalized to an extent whereby 'we find it difficult to imagine otherwise'. It is important then to understand affordance and intuitiveness in CHI as visual, embodied and perceptual in a unified sense. Our motor systems learn to respond to perceptual information in infinitely complex ways and the fact that we may be 'aware' that we are interacting with virtual phenomenon mediated by a screen does not alter the fact that we might feel the process as an embodied whole. Changing a parameter relating to, say, synthesis within a particular VST instrument involves a cerebral action in terms of decision-making, a bodily action through the hands and arms which shifts a virtual representation on screen to ultimately alter the binary code which in turn effects a particular sonic quality which is aurally perceptible to the user. But despite this complex chain of occurrence, the phenomenological experience is instantaneous and undivided. As McCarthy and Wright (2006) note, the sensory is one of the key 'threads of experience' within human–computer relations and 'the body, the senses, and the physicality of the technology are intrinsic' to any form of interaction.

User intuition then is clearly a learnt and internalized process, acquired through an ongoing interaction with a variety of differing applications. In effect, large corporations such as Microsoft and Apple consciously attempt to engender a manufactured intuition by encouraging a standardization of look, feel and function across differing types of application. For example, Apple's *Human Interface Guidelines* provide a comprehensive set of rules for program developers governing the design of an individual GUI. The guidelines exist to explicitly produce a 'consistent visual and behavioral experience across applications and the operating system', the overriding principle being that 'users will learn your application faster if the interface looks and behaves like applications they're already familiar with'.[9] Clearly, it is in the interests of Apple to maintain a carefully constructed set of brand values pertaining to functionality, user-friendliness and aesthetics, but this commercial logic has wider implications in constructing a conventionality of experience across differing types of human–computer interaction.

Such a construction of conformity of feel and function across applications thus has an auxiliary (yet fundamental) effect of re-enforcing the dominant modalities of personal computing upon the practices of digital musicians. All contemporary DAWs are part of the visual logic and functional conventions of the personal computer. The visual organization of what may seem as divergent applications, such as, say, Ableton Live and Logic Studio, is through a selection of windows which serve to visually compartmentalize particular functions such as channel strips, audio and MIDI effects, file organization, etc. In addition, features that are central to operating most DAWs such as drop-down menus, drag and drop, undo functions, file browsers, highlighting (groups of) icons and generic commands are all familiar from

[9]http://developer.apple.com/library/mac/#documentation/UserExperience/Conceptual/AppleHIGuidelines/XHIGIntro/XHIGIntro.html. Bolter and Gromala (2006, 376–7) note that the Apple corporation began to 'commodify' the idea of transparency from an early stage and that such discourses were apparent in the company's interface guidelines as early as the 1980s.

a variety of differing types of application. The intuitive nature of DAWs is therefore intimately connected to the wider modalities of the personal computer itself and the perceived affordances offered by the conventions of the user interface.

From hardware to screen

Nevertheless, the fact that DAWs have a specifically musical function has clearly had a fundamental effect upon the common conventions of their GUIs. The incorporation of the functions of MIDI, sampling and synthesis technologies into the domain of the personal computer meant, quite naturally, that software developers were to some extent initially concerned with creating virtual versions of hardware technology. For example, Duignan et al. (2010) note that there has been an 'evolutionary reliance of DAWs on the multitrack-mixing model' which has fundamentally shaped their representative conventions. Taking a more nuanced tack, Prior (2008a) argues that there has been a gradual separation of the visual feel and utility of DAWs from recording and production hardware. Using Ableton Live as a case study, he notes that in contrast with the first initial wave of virtual studio technologies such as software synthesizers and virtual studios, Live does not even attempt a visual analogue to a hardware equivalent. Instead, the program constructs its own 'aesthetic conventions by creating a "technoscape" comprising relatively distinct modes of working'. In this respect, he places the application in terms of a Baudrillardian simulation (Baudrillard 1988) in that 'its technological form of life refers less to an original referent but to itself, to its own reality' (Prior 2008, 923). Similarly, Bell et al.'s (2015) historical survey of visual metaphor in DAW technology argues that in the emergent decades of digital audio the design of software was dominated by skeuomorphic representation (i.e. visually based upon analogue hardware predecessors). They note that the past decade has seen a trend towards the converse, with hardware being increasingly modelled on software.

However, reading the development of DAWs as a march from simulated and remediated virtual versions of analogue equipment and functionality to a self-enclosed virtual logic tells only half the story. By tracing back through the development of the DAW, we can see that the rather obvious analogue to real-world physical technology was necessarily mediated through the organizing principle of the window from the offset. We can see this in terms of what Bolter and Grusin (2000) term 'remediation': that is, the way that various forms of new media attain prominence by refashioning and paying homage to earlier media forms. In the case of DAWs we can see remediation occurring on two different levels. While early DAWs virtually replicated various physical elements of technologies such as the mixing desk or step-sequencer through on-screen graphics, they were also necessarily grounded in the metaphorical nexus (see Blackwell 2006) that had dominated the development of interface design in the previous decade. It was in this way that DAWs developed their own conventions of representing amplitude, frequency, pitch and time. Hence, DAWs evolved in a particular trajectory in which they standardized their own hybrid visual language through a process of dual remediation. Early programs aimed at both the professional and amateur audio market were often based around the organization of MIDI information, which was used to control external hardware such as drum machines, synthesizers, sequencers and samplers. Even at this early stage, a visual and functional lineage can be traced between applications developed by software companies in the mid-1980s such as Hybrid Arts, Mark of the Unicorn and Mimetics, and mainstream computer-based DAWs of today. Programs such as EZ Tracks, ADAP for the Atari ST, Professional Composer and Professional Performer for the Macintosh, and Soundscape for the Commodore Amiga all used windows-based interfaces and are based on a visual/tactile functionality that would be familiar to any contemporary DAW user. A review of these vintage technologies reveals numerous elements that have become standard within the language of VST: the horizontal bar as the visual analogue to particular types of musical information (bars as blocks, graphic waveforms, 'piano rolls', etc.), file browsing areas, grid-based step-sequencers, zooming functionality, cut and paste and so on.

While to a certain extent these visual representations worked along the lines of metaphor, standing in for the functions of hardware equipment (Théberge 1997, 227), their translation into the virtual at this stage served to render them as distinct. Capturing these metaphors (for time, amplitude, pitch, etc.) within windows for particular GUI systems meant that they were contained in very particular ways and were subject to the graphical limitations and conventions of their respective platforms, creating a new set of abstraction mechanisms (Duignan et al. 2010). Significantly, these programs also utilized a new set of actionable visual elements that drew upon the iconography and generic visual functionality of other key productivity applications. Pointing and clicking, zooming, cutting and pasting, the use of file browsing areas and drop-down menus were all fundamental to how they represented, organized and manipulated sound. Hence, we can identify a dual remediation here in which the adoption of generic visual architecture and functionality, combined with a very particular remediation of the tasks of existing hardware equipment, created a distinct visual template from which the DAW would develop. In effect, the combination provided a new form of conceptual ordering and a set of gestural conventions that would become integrated into modalities of production and composition over the following decade.

For instance, Hybrid Arts' ADAP, the subject of Gary Kildall's comments which started this chapter, was an early straight to hard disk digital recording system released in 1986 for the Atari ST platform which allowed users to sample and manipulate external sound sources. Once a sound had been recorded it would appear on the screen as an audio waveform (a graphical representation of the actual sound in terms of frequency and amplitude). The particular visual interface allowed users to select and highlight sections of the waveform that could then be cropped and altered according to a set of icons positioned just below the waveform window. This visual representation is in contrast to hardware technology that fulfilled a similar function. Equivalent samplers of the time tended to represent captured sounds in numerical terms or through very basic volume bars that fluctuated in a temporally unfolding way as a sound progressed in time. The waveform's capturing of time and amplitude into a static graphic representation provides something different: sound conceptually frozen in time.

In this respect the ADAP was in itself part of a continuum that stretches from the launch of the Fairlight CMI to the present day whereby sound is reimagined visually. Although out of reach to the vast majority of musicians (it retailed for over $25,000 in the United States, Fink 2005), the Fairlight CMI had a foundational effect upon subsequent technologies. The manner in which sound was altered in the Fairlight was in contrast with the knob-and-switch-controlled analogue synthesizers that had preceded it. Aside from the chromatic keyboard which was used to play notes, the Fairlight's functionality was primarily visual. It was the first piece of musical technology to include a screen-based sequencer and graphic representations of soundwaves, and had a waveform drawing page where harmonic profiles could be drawn directly on to the screen via a light pen or particular parameters could be altered causing a subsequent change in sound. The CMI's success among a whole host of high-end professional consumers (including Peter Gabriel, Todd Rundgren, Kate Bush and Duran Duran) meant that it had a significant cultural impact and made it a formative influence on the musical technologies that would immediately follow. The Fairlight CMI is often posited as the beginning of the dominance of the computer in popular music (Russ 2004, 234; Leider 2004, 65; Fink 2005), and subsequent developers were clearly aware of its visual functionality in the process of designing their own equivalent software.

There was also convergence of a number of wider technological factors that would take this template to a much wider audience. The launch of Passport Inc.'s MIDI/2 and MIDI/4 sequencers (for Apple II and Commodore 64) in 1984 constituted the first US PC-based sequencing program on the market. The program had limited functionally in that it could only loop MIDI sequences, notes had to be played into the system by a MIDI keyboard and the system had a rudimentary windows-based interface. Subsequent applications built upon this template, gradually increasing functionality through the addition of layers of windows. The launch of the Atari ST in January 1985 (which included a MIDI port as standard) seems to have been both a sign and a catalyst of the exponential growth in MIDI-based personal computer programs that followed (see, for instance, Mace's 1985 contemporary survey of music software). Jacobs and Georghiades

(1991, 31), for example, point out that although the Atari ST was launched principally at the games market, its MIDI and disk drive capabilities, and its 3-channel sound chip made developers realize its enormous potential for music making.

It is also crucial that these programs were emerging at the same time as concurrent developments were taking place in computer operating systems. The release of Apple's Macintosh in 1984 was the first commercially successful computer to use a GUI and adopted the visual metaphor of the desktop with 'folders' and 'windows' that remain to this day. Hence, early computer-based DAWs were being developed for operating systems which had windows as a fundamental organizing principle of their GUIs from the very beginning. It is therefore quite understandable that there was a conflation of the representation of musical concepts and common set of visual metaphors used in these windows-based GUIs. The specificities of the historical context in which DAWs evolved have meant that the embedded visual metaphors of computing have been written into the DNA of their evolutionary trajectory. Thus the 'scripts' (Akrich and Latour 1992, 59) provided by the DAW that mediate action should be understood as simultaneously encompassing musical and (traditionally) non-musical systems of organization. However, what I want to suggest here is that through a sustained period of enculturation, the more general organizing principles of the GUI *become* ways of structuring musical thought within the creative process.

Native DAWs and multifunctionality

We can see the half-decade or so between the Atari ST's launch and the early 1990s as a period of rapid naturalization of the DAW as a personal computer-based concept (as opposed to the integrated hardware systems that had proceeded it). In these early examples we can see the gradual standardization of the visual representation of audio tasks leading to a shifting modality in terms of how sound is thought about within the creative process. The enculturation of DAWs as a norm within musical creativity across a variety of types of musical practice (from electronica to mainstream pop production) imposes a broader conceptualizing of sound in visual form and its manipulation through a hand to eye coordination.

As DAWs progressed, they began to gradually integrate the functions that had previously been the domain of the hardware they controlled, leading to a tightening of the action-perception loop between screen and gesture as the screen/controller became the main focus of user activity. This was largely a result of the widespread availability of faster processing speeds which were increasing incrementally throughout the 1980s. Increases in speeds meant that processors were able to simultaneously handle audio processing as well as MIDI. By the mid-1990s native DAWs began to emerge which included

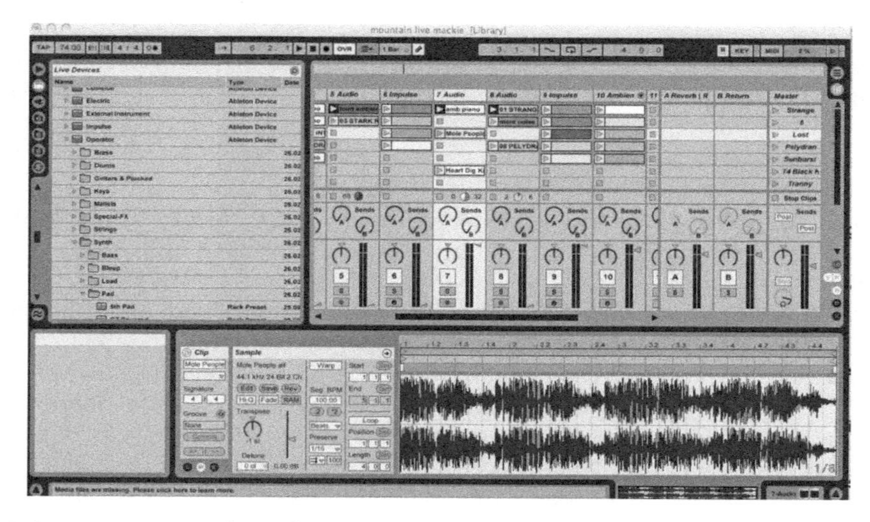

FIGURE 2.1 *Screenshot of an Ableton Live session. Live's use of 'clips' as the primary unit of creativity treats MIDI and audio as concomitant types of musical information.*

audio recording and manipulation, 'soft' synthesizers, step sequencing, audio and MIDI effects all within one package. The evolution of different versions of long-standing DAW software brands has thus seen them becoming more uniform as they take on differing tasks. Logic and Cubase had initially been launched in the late 1980s as purely sequencing and MIDI applications, but progressive upgrades saw both programs incorporating audio recording and editing facilities in the mid-1990s. Similarly, Pro Tools evolved as a hard disk digital recording package, only introducing MIDI capabilities in 1999. By the time that Ableton Live 1 was released in 2001, there was a growing expectation within the market that everything in the production process could now be done 'in the box' within one multifunctional application. Industry statistics point to a significant change in usage within the electronic music market during this period. The decade between 1997 and 2007 saw a huge rise in the value of the musical software and soundcard-related products while the same period saw a severe slump in both hard disk/cassette-based multitracks and stand-alone sound modules such as samplers, integrated hardware DAWs, step-sequencers and drum machines (NAMM 2007, 27–30). In the United States for instance, sound modules went from being worth \$45 million in 1997 to below \$10 million in 2006 (ibid.) This crash within this particular sector of the market was mirrored by an increase from \$12.75 million to \$73.7 million in the dedicated soundcard market in the same period.[10] Clearly, these statistics suggest that for many

[10]Piracy: what do the numbers say http://www.imsta.org/piracy.php

consumers the computer itself was becoming the core focus for musical production and creativity. MIDI composition, recording, editing and mixing could now be done in one environment, clearly a significant development in how the creative process is imagined as a whole. As the preceding chapter has illustrated, even in the case of heavily funded popular music recordings there are ever increasing instances where most of the production process has been completed within a particular DAW. The Moroccan producer RedOne describes his productions with Lady Gaga as '100% Logic Studio' even to the extent of a sole reliance on its built-in VST synths, drum modules and effects units. He comments that even though spatially the Lady Gaga records were recorded in high-end studios, much of the audio work was being done within his own computer: 'All the hits I did with Gaga, honestly, it was a funny thing. We were working in big rooms [studios], but we were using my equipment. Like my Apple studio speakers, and we were working from my laptop most of the time' (quoted in Levine 2010).

The incorporation of distinct production tasks into singular applications meant that DAWs were no longer singular parts of a chain within a MIDI studio set-up[11] but native computer programs which had developed their own conventions in accordance with the wider historical development of personal computing over a significant period. As Prior notes, this is particularly significant as by the 2000s 'a new generation of musicians with little or no experience of the hardware studio is learning to make music in software environments, displacing the referent [mixing desk, sampler, keyboard, etc.] further from its image' (Prior 2008a, 923–4). In short, for early DAW users, the idea of virtuality was just that: the screen-based interface provided an analogue to a particular process associated with a real-world piece of equipment. After all, most professional and hobbyist users in the 1980s would have had at least a passing familiarity with such hardware. In contrast, computer-based native DAWs engender a type of thinking about musical information in which MIDI and audio are seen as two facets of the same process rather than disconnected elements of separate hardware. Again, Ableton Live's ordering of musical information is instructive here. The program's primary unit of the 'clip' does not make any distinction between MIDI and audio at the basic interface level. Both are merely facets of musical information that may be processed in very similar ways once that information has been dropped into the program's main window (see Figure 2.1). In effect, Live's particular set of visual affordances can be seen as the ultimate fruition of the hybridizing processes that had been going on in DAW design since the 1990s. In addition, a generation of musicians and producers had also grown up with the functional norms of personal computing across their daily lives, from work to leisure. Visual and conceptual elements that

[11]Daisychaining (or the linking of devices) was a central part of how MIDI was utilized up until the domination of computer-based DAWs.

had become central to the design of DAWs, such as the spatiality of the window, database thinking, drop-down menus, dragging, dropping, cutting and pasting, were for these digital natives (Palfrey and Gasser 2008) a logical way of ordering information per se. The implications of this naturalization upon the composition and production of music are profound, in that the imbrication of the everyday with the creative has led to a restructuring of how musical materials are thought about and put into action. The digitalization of composition folds everyday modalities into the DNA of the creative process. For example, Prior argues that 'as code has become a visual representation of a studio, with its simulations and icons, so the user has been presented with a surface on which to skim and play. Writing music in this way constitutes a flexible practice, subject to the speed of a copy/paste key combination or undo stroke, while the interface represents the work as a malleable digital landscape' (Prior 2009, 87).

Summary

Returning to the vignette with which I started this chapter, the feeling of intuitiveness that characterized my own experience of a particular creative encounter can be understood as situated within a complex unfolding of historical and technological circumstances. My ways of working were partially structured through a conceptualization of sound facilitated through the visual representation of the interface of that particular program and also what I had internalized in the longer term through my use of other DAWs for the decade leading up to that point. My bodily and visual engagement with the interface was a result of my own tacit construction of a set of creative working practices, but it also took place within my computer use more generally. Many hours of each day of my life are spent interacting with computer screens for a variety of tasks, from teaching my students with Microsoft PowerPoint to word processing for research and administration, from online shopping to checking out the latest music releases, from communicating with friends to contacting my bank. Each of these activities and the programs used to facilitate them might be seen to produce differing types of experiences and feelings and each might construct a differing set of modalities. Nevertheless, they are all connected through a shared evolutionary strain. Their modalities are connected through a shared design logic related to discourses around affordance and intuitiveness that have been within the mainstream of the computer industry since the 1980s. Hence, for me, my creative practice in this instance was part of a wider interactive engagement that permeates a large portion of my waking life. That wider engagement has the capacity to be boring or exciting, passive or highly engaged. The heterogeneity of experience engendered by our engagement is precisely the point; it is a result of the sedimentation of knowledge and a subsequent perceptual transparency.

The entwined design logic outlined here intersecting with personal, individualized uses of technology also has implications for our understanding of the nature of computer–human relationships in this context. For example, Zagorski-Thomas (2014) applies Akrich and Latour's (1992) notion of program and antiprogram to explain human actors' relationship to inherent scripts built into technologies: the 'inscription by an engineer, inventor or manufacturer' (Akrich and Latour 1992, 259). He argues that 'there are usually multiple affordances but only one "correct" script designed into technology' (Zagorski-Thomas 2014, 15) and that usage might be in conflict with the script can be understood in terms of the antiprogram: that is, 'the programs of actions of actants that are in conflict with the programs chosen as the point of departure' by the designer (Akrich and Latour 1992, 259). This perhaps gives a too rigid formulation of computer–human relationships in this context. The embedded intuitiveness resulting from the DAWs' specific trajectory of emergence within wider discourses that pervaded the entire software industry serves to problematize this rigidity. While, of course, DAWs are designed with specific tasks in mind (and in that sense 'inscripted'), in many cases the design logics encourage exploration, personalization and play. In effect, users are encouraged to find their own uses and to adapt the functionality of the software towards their own ends. Rather than merely providing goal-orientated outcomes, these digital artefacts are active in the 'staging of experiences' (Bolter and Gromala 2006) for users, which are more fluid and open-ended.

The design logic of DAWs and their particular affordances can be clearly linked to the changes in production technologies outlined in the previous chapter. On the one hand, we can see this 'democratization' as broadening the amount of individuals with access to information and technology that was hitherto the domain of specialized professionals within the recording industry. On the other, the widespread availability of DAWs serves to produce new types of knowledge and new types of creative individuals. As the historical outline within this chapter has suggested, the functions of DAWs are not simply a virtual version of the recording studio or hardware MIDI sequencer. Rather, their particular combination of audio processing and MIDI produce distinct modalities in terms of musical creativity. In other words, DAWs can be seen as hybrid technologies that act in the construction of a new set of creative conventions and ways of working. As the next chapter explores in more detail, they have had material implications with regard to what electronic musicians actually do.

CHAPTER THREE

Digital technology and technique in the creative process

Mountain, Liverpool 2010

In my project studio I am about to begin work on a new project. I have been asked by my friend, the pianist Anni Hogan, to work on an album she has been working on. After a series of initial discussions about how we both envisage the collaboration, Anni has given me a DVD of recordings of partially improvised piano pieces made at Liverpool's Parr Street recording studios in the previous year. It was agreed that I would use these recordings as a basis for a series of new pieces that will form the core of a full-length album for the British industrial/dark ambient label Cold Spring.

I do have something of an institutional and personal framework which might guide how the project will eventually sound. On a personal level I want to do justice to Anni's nuanced and atmospheric playing. Through my discussions with Anni, we have decided that we want the project to be experimental and 'dark' in its overall feel. In addition I have an implicit idea of the record label's usual sound (it is most well known for releasing music by acts such as the Japanese noise artist Merzbow and some of the best known names in industrial music such as Coil and Psychic TV). Nevertheless, the project is something of a blank canvas at this stage, certainly in terms of the sonic nuances of the finished product.

Initially, I drag the files across directly from the DVD window directly into Logic Studio's arrange window and begin cutting up the various sections of one of the audio tracks. After sometime looping and reviewing the cut-up sections, I have identified distinct sections that might work together but nothing is giving me an immediate sense of direction of where the piece might go. I decide to change tack by changing computer program. I bounce down the sections I have identified into new audio files and open Ableton Live. I drag the new files into a clip slot within Ableton Live's graphical user interface and begin the process of editing and looping. I double click on the file section of one of the clip slots which opens a sample editor towards the bottom of the screen displaying a graphic representation of the sound file. I warp and loop the file, an action that automatically renders it in time with the overall tempo I have set for the track, and over the next couple of hours I select differing sections of the audio material using a mouse. Through a process of looping, dragging the 'loop brace' across the waveform in often arbitrary ways and aurally reviewing the results I eventually settle upon a series of sections that will form the basis of one of the album tracks. This constitutes the first stage in what will become a process of months of editing, experimenting, playing, structuring and mixing before the project is finished.

At every stage of the process I use a similar set of techniques to experiment through trial and error before settling on a particular feel or structural trajectory for each track. I will drag and drop, chop, reorder, loop and zoom in on the files before dropping various types of audio effect on them aurally reviewing as I go. Within this process, the sonic elements within a particular recording might suggest certain directions for a track or it might be a particular layering of effect on to a small loop that hints at a way forward. In some tracks the piano pieces will be framed with minimal processing: a reverb here, a splicing of audio there, and their melodic and harmonic progression will be left to speak for themselves. In others the piano recordings will be treated as objects which throughout the process will be spliced, time-stretched into new structures, which will maintain only echoes of the original played performances. In short, for this project the original tracks are taken less as 'piano pieces' than as sonic source materials that will be utilized in varying ways. The musical affordances apparent in the original tracks are taken in a variety of directions – sometimes according to their directly 'musical' properties (pitch, melody, etc.) while in others they are taken on a purely textural level whereby particular sonic qualities become a starting point to stretch out the material into new shapes and terrains. Often this 'materialist' approach to composition will take the audio material to places not immediately apparent from their original form. Once these more abstract forms have taken shape I am not entirely sure how I reached a particular point. My level of engagement has been such that new structures and textures seem to emerge without any particular awareness or memory of a clear trajectory in terms of process.

Affordance and process

Like my performative experience in Glasgow offered at the start of Chapter 2, reflecting on the beginnings of what would become a long-term project (the *Mountain* album would not be released until 2012) raises a number of issues relating to the experiential properties of my own interaction with digital technologies. Partially, these are pragmatic issues of learning and personal working habits. I was approaching a new project by dealing with sound in a number of set ways of doing things that I have tacitly acquired over a number of years. Yet the levels of absorption and mental flow that I experienced perhaps reveal how these personal working conventions are part of a more fundamental shift in my musical thinking. The techniques

I used in working towards a musical end product have, for me, become more than just techniques. They are ways of conceptualizing sound and how sound relates to wider fundamentals of music. In the context of digital production, rhythm, structure, timbre, feel and signification are all connected with a set of processes that I have used and reused until naturalized. I have an understanding of the ways in which various types of audio processing or effect can fundamentally shift the feel and mood of an existing piece of audio. I use MIDI information as a rhythmic or melodic skeleton on which timbral elements can be overlaid, tried out, reviewed and altered. I have an idea of how the kernel of an idea can be gradually built, layered and stripped, before finally creating an overall musical structure. All of these elements are at the same time situated within, and used as expressions of, my wider musical consciousness: what I have internalized about musical 'rules', aesthetics, genre and the affective power of music. Sitting down in front of a computer screen there is no disconnect between the conceptual and the processual: they are experienced in chorus as one involved encounter.

These personal reflections are again indicative of how digital technologies have a material effect upon our consciousness and subjectivity at particular moments in time. Technologies are active in what we do and also in the way that we feel. And it is this production and negotiation of experience *through* and *with* technology that forms the basis of this chapter. It argues that techniques related to specific technological environments are intimately connected with ways of being creative and ways of thinking about musical creativity for the digital musician/producer and seeks to examine the specificities of those techniques, connecting them with common conceptual ordering.

In order to unpack these issues there is a further concentration on DAWs throughout the chapter. In particular, the main purpose here is to explore how these technologies operate as an actor within the creative processes. This is not to suggest that technologies determine the creative process. Rather, I am interested in the way in which digital objects define a broad framework for action but are subject to their intersection with human actors and agencies in a 'process of exchanging competencies, offering one another new possibilities, new goals, new functions' (Latour 1999, 182). The previous chapter outlined some initial ways of thinking about DAWs in a connected historical continuum. It suggested that they are audio-focused applications which simultaneously rely upon the canonical affordances central to the broader conventions of GUI design that have been developed in the computer software industry since the 1970s. This historical account was used to suggest that DAWs engage sight and touch in ways that have become increasingly enculturated into how we undertake a variety of computer-based tasks in everyday life. Thus, the very specific historical development of digital music technologies can be seen as shaping how sound is perceived for the creative individual.

With its concentration on the experiential, the chapter in hand examines how the creative process actually unfolds within this context in more detail.

A key argumentative thrust of this chapter, therefore, is that the development of visual affordance of the type outlined in the previous chapter has had a clear effect upon how musical affordance is framed, and how certain ways of thinking about music are constructed. It is specifically concerned with how the experiential properties engendered by the DAW mean that the visual and the sonic are in a constant state of overlap. What I want to suggest is that the simultaneous apprehension of visual and sonic affordance creates connected ways of conceptualizing, ordering and working with musical materials. Thus, the chapter's concern is with how the affordances relating to human–computer interaction outlined in the previous chapter combine with musical affordance within the creative process.

In order to unpack this connection, the concepts of musical and sonic affordance are introduced in order to sketch out the theoretical basis for the remainder of this book. It is suggested that there are certain sonic and musical properties that may be apprehended within the materials (such as recorded sound and synthesis) that provide the prima materia of making music in a digital context. These sonic affordances provide routes into possible creative action for the musician/producer. The chapter then goes on to examine how these sonic affordances are always situated within distinct modalities of creativity related to the visual and metaphorical conventions of HCI. The chapter then makes some broad observations about how spatiality is constructed by DAW interfaces, before examining a series of functions and organizing principles common across differing DAWs. As examples, actions such as freezing, layering, drawing, slicing, looping and zooming are explored as both functional and conceptually instructive.

Musical affordance

In its application to music, affordance theory began to be used in the 2000s in two important ways: first, within reception studies relating to Western art music and secondly, as providing a valuable way into thinking about the varying uses of music in daily life more generally. In both of these cases, affordance has been applied as a theoretical tool to conceptualize music as a material that is active in structuring embodied or cognitive experience. For example, work such as De Nora (2000) and Kreuger (2010) has been generally concerned with how music is put into action in the service of identity, interpersonal coordination and emotional regulation. For scholars such as Windsor (2000); De Nora (2003); Clarke (2005); and Nussbaum (2007) the concept of musical affordance has been a useful way of addressing ongoing musicological debates around issues of musical connotation, representation and use while avoiding any reductionist tying down of the meaning of a given text. The basic premise that cuts across both of these types of work is that music contains within it certain actionable properties that are put to work by a perceiver in a number of ways and according to a number of social and personal factors. This relational approach has certain commonalities with the application of semiotic theory to music but allows for a move beyond a text-centred approach that circumnavigates the rather knotty set of issues around musical meaning.

Within the fundamentals of semiotic theory a sign is understood to 'stand for' something else and as being 'representational' of a signified other. Within music this is seen to occur through a mixture of repeated use within specific cultures, paramusical and extramusical association (for the relationship to other musical texts and non-musical sound associations respectively, see Tagg 1991). It is in this way that musical signs are seen to operate within a broader socially constructed context of shared understanding (or the continuum of signs that make up a 'semiosphere' Lotman 1990). While affordance theory does not directly contradict the underlying premises of semiotic theory (in that it acknowledges that music has signifying and affective power), it brings the relational connections between the sound and perceiver to the centre. It should also be noted that the relational is implicit within musical semiotics. As Dunsby (1983, 28) notes, 'For semiotics, music is a cultural or social phenomenon, definable only in terms of its value held in a culture according to a quantative, qualitative and analytical interplay.' The key difference is affordance theory's foregrounding of experience, perception and, critically, use, a legacy of the term's basis in ecological psychology. The concept of musical affordance thus foregrounds the 'relationships between, rather than meanings of [the] cultural elements' of music (DeNora 2003, 46). The presumption that there are certain qualities that present themselves within a musical object which afford action or response by the perceiver has broader implications as to what music 'contains' and how it works in the world. Rather than having inherent qualities that signify or have tied meanings, music has qualities that are brought into use. This concentration on use therefore allows for the possibility of a plurality of meanings, employment and social function 'without being swept away by total relativism' (Clarke 2003, 119).

For example, Nussbaum's (2007) approach is to treat representation within Western tonal music as a Gibsonian surrogate (or external representation) that operates in a comparable way to a painting or a moving picture. Music is thus active in the representation of 'virtual objects' and is 'informationally structured' in particular ways (in terms of motion, space, etc.). This position should not be misunderstood as another of way of simply saying music has meaning in a deliberate critique of the absolutist disavowal of musical meaning external to the text. Rather, it is a way of viewing properties within music and how they operate in the world. Clarke (2005) notes that musical affordance is always mediated through the social, and that differing levels of knowledge, experience and socialization lead to differences in how music is perceived and placed within a particular framework of understanding. He outlines how different individuals have different 'perceptual capacities' as well as differing sets of musical values and experience. An underlying assumption in Clarke's overall argument is that auditory information necessarily has particular affordances but how they are perceived is dependent on a number of social factors.

Similarly, for De Nora (2000, 2003), the concept of musical affordance provides a 'middle way ... between the analysis of music as discourse and music as action' (2003, 154). In other words, both text and perceiver have agency with regard to music as being affective and meaningful. Musical affordance suggests a complex contextual framework through which music is always mediated, and how music is 'drawn in to action' for these ends (ibid.). These include preconditions, that is, particular discursive conventions, biographical associations and other contextual factors relating to musical culture and reception. These preconditions intersect with what she calls the features of the musical event. A musical event may be of any duration but features actors (the individual who is engaging with musical material), music itself, the act of engagement (i.e. what is being done) and local conditions and environment (i.e. how did the individual interact with that music in that particular way and in what setting did the musical events take place). It is according to this framework that perceivers use music to construct a particular sense of identity or community, to chime with or regulate their physical surroundings, mood or emotional state.

Kreuger (2010) makes an even more fundamental claim for musical affordance by suggesting that it can be related to elements of listening which are deeply embedded into our perceptual wiring. He uses research into neonatals' engagement with music (Standley and Madsen 1990; Schellenberg and Trehub 1996; Trainor and Heinmiller 1998) to suggest that, from birth, music is perceived differently from other noise, and that various forms of socialization mean that music is a central part of the entrainment (i.e. the coalition and association of bodily features to habitual features within the environment) from an early age. For example, he notes that there are certain elements within music that invite appropriation even by babies. Simple textures and predictable melodic and rhythmic patterns, for instance, provide a soothing function which the infant can utilize and master and are thus musical affordances which 'invite the infant to attune itself to its sonic world' (Kreuger 2010, 15). In other words, there are affordances that are congruent with the physical; with bodily action and rest, with longer perceptual rhythms and cycles, through which music can be instrumental in anchoring a sense of mood, place or emotion. Despite the rather elemental thrust of this argument, Kreuger goes on to note that musical affordances are far from static and are often clearly rooted in the social. He gives the example of the communal experience of music and how it engenders a collective attention arguing that 'a mutual attunement to the social affordances in music modifies how music is given (i.e. phenomenally manifest) to multiple experientially integrated, perceiving subjects' (2010, 17). In other words, the act of being within a particular physical proximity to others within a particular socially prescribed setting alters the way in which music is afforded and thus the perception of the individual along collective lines.

What is particularly useful within this work in terms of understanding musical creativity is the position that sonic materials have a particular suggestibility which perceivers make sense of according to particular nuanced contexts. These qualities can be both fundamentally linked to how the body is positioned in the world (in terms of pulse, rhythm, spatiality, etc.) and subject to complex patterns of social construction. Affordance is thus a way into thinking about how sonic material intrinsically presents musicians and producers with what that material could be used for within a particular artistic framework. Affordances provide invitations to action according to attributes that are already present within musical material. Just as a listener may perceive a musical affordance and put it in to a variety of actions related to meaning, feeling, mood or sensuality, so the creative individual may perceive a number of actionable possibilities in a sound, melody or rhythm.

Sonic affordance

In light of this, it is perhaps pertinent to make some kind of distinction between musical and sonic affordance. In all of the work discussed here, the idea of musical affordance is ultimately reliant upon a perceiver's understanding of a set of organized sounds *as* music. Musical affordances may have underlying invariant properties that are externally representational or chime with a particular physical or cognitive disposition, but they are nested within a distinct structural framework. As Krueger (2010, 5) notes, musical affordances are realized 'within the relation between a feature of the environment (e.g., particular structural qualities of a piece of music) ... and a perceiver-side ability or skill (e.g., motor capacity, perceptual, and affective sensitivity) enabling the pickup and appropriation of this structural feature'. For Kreuger 'the relevant perceiver-side skills are what make music affordances show up as available for engagement and appropriation' (ibid.).

Sonic affordance, on the other hand, has been used more generally to refer to how sound has actionable properties per se, outside of the structural context of music. Particular sounds have distinct physical qualities that are perceived as meaningful or actionable depending on where, and by whom, they are experienced. Oliveira and Oliveira (2002, 8), for instance, give the example of a runner on a starting block waiting for the sound of a starting gun. The sound affords the action of running in order to compete in a race in the quickest possible time, whereas if the same sound were heard in a bank or a classroom, it would afford searching for protection or flight. Sonic affordance also operates at a more mundane level, from the buzz of the morning alarm to the tranquil bubble of nature in an urban park, from the pitter-patter of rain on a window to the laughter or cry of an infant. In other words, the relationship between sound and perceiver is constantly locating us within space and time, presenting us with thousands of minute choices every day. Sonic affordance is thus present as a process across our daily lives and is realized independently of musical structure.

By making this distinction I am not claiming that the two are in any way mutually exclusive. Sonic affordance can be translated within musical material, and sonic and musical affordances are constantly overlapping. However, there is clearly a difference in terms of the use of recorded sounds within music as they are necessarily abstracted from a real source in actual proximity through their mediation. In a discussion of electro-acoustic music, Windsor (2000, 17) makes a distinction between real and virtual sonic events. While both are apprehended in the same way, through a certain invariance in terms of their physical acoustic properties, the perception of virtual sonic events is 'non-veridical', in that they do not inform the perceiver about their real environment. Nevertheless, listeners still 'exploit their sensitivity to invariancies in individual sounds or the structural relationship between them' (2000, 18) during the listening process. Similarly, Tarasti's (2002, 49) application of a Piercian semiotic model suggests that the uses of 'noises' in music are clearly examples of 'first articulation' in that they 'have a certain denotation on the basis of their recognition'. In other words, the 'real world' relational properties of a sound are clearly apparent within listening. This is not to say that once a sound is mediated through recording it carries with it the same set of affordance structures, rather that they are implicit even in their mediated form. If everyday sound carries spatial and actionable properties, when such sound is appropriated within music it carries with it the residue of the relational aspect of such sonic affordance.

The implications of this difference have been the subject of a central debate within electro-acoustic music since Schaeffer's original demarcation of the acousmatic as a fundamental principle within the apprehension of recorded sound. Schaeffer (1966) proposed a conceptual framework for compositional practice whereby the composer should treat sonic material as a sonic object (*objet sonore*), which should be apprehended through a reduced listening. In this context sound should be valued for its own sake, its intrinsic sonic qualities, stripped of its source and conveyance of meaning. The aesthetic implications (and impossibilities or otherwise) of Schaeffer's theorization have been discussed in great detail elsewhere (Wishart 1986; Holmes 2002; Emmerson 2007; Kim-Cohen 2009) and it is not my intention to rehash these debates here. However, Schaeffer's empirical kernel of the sonic object is, I think, pertinent in raising questions about how we perceive pieces of sonic information. Schaeffer's core concern was with the fact that underneath associative and actionable relational properties of a sound, there are certain sonic qualities of acoustic grain and texture that are fundamental to how it is experienced. In other words, there is an immediate apprehension of a sound in terms of its materiality that intersects with our physicality. It is through the materiality of sound that our interaction with sound objects solicit affective response in a similar way to Massumi (2002) and Brennan's (2004) identification of 'autonomous affect' to describe sensations afforded by media or environment that are presocial or non-linguistic. We can understand affect in this sense as describing a neurological response to external stimulus or an independent bodily reaction when a perceiver is confronted with a particular phenomenon. Meelberg (2009) uses the term 'sonic stroke' to describe the immediate sounds within music that induce a frisson through their volume, timbre or rhythm and have an effect upon the listener's body.

Sonic/musical affordance and the creative process

These differing approaches to the affordance structures of sound and music illustrate the complexities of creative decisions made by electronic musicians and producers. Sonic materials may afford particular creative uses through their materiality, their perceived tonality or their real-world associations. In actuality, a singular sound object offers a complex overlapping of perceptive, bodily and associative possibilities. For example, a sample taken from a previous musical recording is immediately, and simultaneously, perceived according to differing properties. We identify the sonic information as music but also perceive its physical and textural surfaces, its spatiality, its frequency, its actionable properties.

Despite the slightness of this distinction I do think that it is one worth making for the purposes of this book. Because digital musicians/producers are working in an essentially timbral tradition (Hugill 2008), sonic/musical affordances and the immediate affective properties of sound are perhaps more pronounced than in other forms of musical creativity. Working with DAWs often involves working with pre-existing audio materials, be they found sound, instrumental samples, field recordings and other self-recorded files or samples from other musical recordings. In addition, the use of MIDI-controlled synthesized sound and the manipulation of audio effects form the other core stems of creative practice for the digital musician/producer. Both similarly require a constant assessment of sonic surface and texture during the creative process. In both, the affordances and affective qualities of a particular sound, be it a synthesized tone, a recorded ambient hum or the partial harmonics of a particular reverb setting, can spark off a particular creative trajectory. In contrast to, say, the composer working on stave paper with explicit procedural knowledge in mind (in terms of melodic or harmonic theory, conceptions of the structural conventions of differing types of musical 'work') or the singer-songwriter fishing for ideas with her acoustic guitar (which will later be articulated in differing instrumental arrangements during performance or recording), composition in the digital realm is laden with sonic affordance. To a large degree, from the very beginning of the process digital musicians are working with what is already there, with sounds that already exist (and ultimately sounds which may go on to exist within the final recording or performance). Thus, the creative and compositional process for these cultural practitioners is to a large extent conceptualized through sound.

In Demers (2010) work on the aesthetics of various types of experimental electronic music she identifies a 'materialist' approach to composition in genres such as electronica, dub and microsound, where sounds are viewed as objects. She notes that the appropriation of existing sounds can be framed and articulated in a number of differing ways according to distinct aesthetic positions. In the sampling practices of artist such as Matmos, John Wall and Steve Takasugi, any connotations of the original sounds utilized within a recording are maintained but are played upon, deconstructed or abstracted in a process of associative play creating new meanings and sonic or affective qualities (2010, 59–63). In contrast she identifies a 'base materialism' within the microsound of Alva Noto and Ryoji Ikeda, a preoccupation with the intrinsic sonic quality of a given sound which serves to 'strip sound of its associations, leaving only the sound itself' (2010, 79). Thirdly, Demers argues that drone and noise music employ elements of the sonic spectrum that are counter to customary expectations of musical language by drawing attention to the physicality of sound that music 'usually asks us to ignore' (2010, 91).

Despite these differing aesthetic positions, appropriated sound fundamentally underpins all of these traditions with sonic affordance, offering divergent creative paths. What these examples deftly illustrate is how sonic and musical affordances do not dictate a creative trajectory for the musician/producer, rather that they are always positioned in relationship to the individual's wider social and personal trajectories. As Windsor (2004, 181) points out, 'Objects and events may afford different things depending upon the needs and affectivities ... of the perceiving organism ... affordances themselves are culturally relative, and open to social mediation'. In terms of musical creativity therefore, factors such as genre, institutional context and personal artistic trajectory form an important prism through which sonic/musical affordances are acted upon.

Monitoring the sonic

While these social factors are dealt with in detail in the next chapter, for now we should bear them in mind as part of a contextual amalgam that constructs and mediates sonic/musical affordance. What I want to concentrate on for the remainder of this chapter is the way that technology (with a further concentration on DAWs) provides a structuring of the experience of musical creativity. In particular, I want to suggest that DAWs provide an environment whereby visual and sonic affordance overlap. On the one hand, DAWs provide an intuitive visual environment in which the affordances of sonic material can be perceived, tested and reviewed quickly and efficiently. On the other, visual affordances serve to reconstruct thought about the conceptual ordering of music within the creative process.

This central positioning of the visual in the creative process is somewhat contradictory. There is the issue that the visual nature of the interface of a DAW might actually distract from the aural aspects of composition and divert the individual's ear from what is really important in the creative process. In Gall and Breeze's (2005) work in schools (with the programs Cubasis and Ejay) for example, they found that children were often guided by the visual, choosing samples because they had interesting names and deciding where to place sounds within the composition according to the visual aspect of the screen without reviewing it aurally first. The findings of this research are echoed in a similar study by Reynolds (2008), whose interview material with children using the DAW software found that their conceptualization of sound was grounded in the visual affordances of screen-based technologies rather than any understanding of musical composition. While there is clearly a difference between schoolchildren working with simplified versions of standard DAWs[1] for the first time and the working practices of experienced musicians who have a wealth of accumulated technological and aesthetic knowledge, we can identify some common tensions in the DAW's particular integration of the visual, aural and conceptual.

For example, there is often a schism in the perception of musicians between the GUI of DAWs being a useful tool that helps to facilitate a smooth and efficient way of working and a medium that imposes an overriding and dominant visual conception of sound.[2] For instance, the UK electronica artist Four Tet comments that:

> People who make music on computers don't realise how powerful the visual element is. Whether you like it or not, your mind starts to think in terms of patterns, because it's a natural human way to do things, and you start seeing the way drums are lining up on the screen, and it becomes completely instinctive to line them up in a certain way. It's important just to close your eyes and use your ears, and trust what's coming out of the speakers more than anything. (Quoted in Inglis 2003)

---

[1]The pupils who formed the basis of the study were using Cubasis and Ejay.

[2]See, for instance, the responses of studio working in Williams (2012).

Similarly, the Danish producer Trentmoller talks about consciously going back to outboard analogue equipment for his 2010 album in order to facilitate a different working process away from a reliance on 'in the box' virtual instruments and visual monitoring:

> Often while recording I was turning off my computer monitor, so I couldn't graphically see what I was playing. Normally when you are working with a computer, you are very locked to this visual image of making music, so it was really great for me to just play. In a way, forcing your ears to be more trustful. (Quoted in Ridge 2010)

These examples are indicative of the way in which digital musicians and producers must constantly negotiate between the visual and the aural in monitoring their creativity. As sonic artists, musicians and producers are acutely aware that sound is the most important facet of what they are doing and that the visual should be regarded as a means to an end. The deliberate minimizing of the DAW screen has become common practice within digitally based studios. In my own observations in various types of studios from project studios to larger, more commercial set-ups, the studio process has been regularly punctuated by moments of purely aural monitoring.

However, it is not my intention to suggest an overriding domination of the visual to the detriment of musical creativity. Rather, the common concerns as expressed by Four Tet and Trentmoller are simply a by-product of the ways in which new modalities have emerged in which visual and sonic affordance converge and overlap. Both have become central to the working practices and hence creative processes of digital producers and musicians. In actuality, there is little contradiction between the two. The creative individual may alternate attention between interface and sound as needs demand,[3] or more often, sound and sight are experienced simultaneously in an unproblematic fashion. This constant need to track back to the musical material itself is also part of a wider tension within the creative process precisely because of its situation within a technological human network. In this way, digital musician/producers are always simultaneously working *with* and *against* technology. They are acutely aware of not only its possibilities but also its limitations.

[3]For example, Bates' (2009) fieldwork in Turkish (Pro Tools-based) recording studios outlines the way in which assessment of, and engagement with, the production process is facilitated by a variety of tactile, sonic and visual manoeuvers which are integrated into the working practices of sound engineers and musicians without apparent contradiction or crisis. Visual assessments of bass frequencies or the shape of overall mixes were seen as vital in terms of the production process without being seen to override or unduly effect purely sonic monitoring.

Modality and creative action

DAWs not only provide tools to realize the potential of sonic affordances, they also provide a structuring framework that has an active effect upon how these creative processes unfold. In their research with professional record producers, Duignan et al. (2010) found that 'the majority of work being conducted within the framework of various DAW systems, where the developing composition is held in an abstract representational form, and participants' abilities to manipulate that representation were entirely mediated by and dependent on the mechanisms provided by these tools'. In other words, the GUI of the DAW constitutes not only the nexus of creativity for such musicians and producers but also a way of perceptually ordering sonic material and conceptualizing sound synthesis. Within this context, the simultaneous apprehension of visual and sonic affordances creates connected ways of conceptualizing, ordering and working with musical materials. In terms of the experience of creativity, the combination of the familiarity of layout, the apparent naturalness of hand-eye perception-action loop and a tacit assessment of sound, offer a connected set of complex affordances (Turner 2005) for the producer/musician. Here, sound is both heard and simultaneously conceptualized within a visual and metaphorical matrix.

In many cases, when a musician begins working with a new piece of equipment, the embeddedness of intuitive logic within the working practices of designers means that the layout of the new GUI often suggests particular uses through its spatial organization. This constitutes a kind of 'in-scription' (Akrich and Latour 1992, 259–61) of intuitiveness within the surface of the digital object. For example, in a promotional interview for the audio equipment company Native Instruments, the jazz musician Kevin Sholar talks about working with the company's hybrid MIDI controller/software product Maschine as part of a collaboration with techno DJ Carl Craig.

> It's so easy to make beats right off the bat. I didn't have to spend six weeks understanding the GUI. I just had ideas, sat down played the pads and it worked out. Including a lot of the songs I had worked out in Logic and Live, when I sat down and worked with Maschine it made my workflow easier, thinking in the concepts of scenes and patterns. [It was] more thinking in terms of Ableton Live for creating music with loops than thinking like I used to think with Logic having these whole complex arrangements. I like the idea of how you start with a pattern and then the pattern combined with a group creates scenes. To me that's a two-dimensional concept. Whereas you have one component that is just rhythmical and another which is based on sound groupings. And I really like that because sometimes I like to take patterns and try different sounds … and it made it easier for me to get between different types of

> sound without having to re-program everything ... not just in the studio but also live to be able to see and work spontaneously.[4]

Sholar's comments are illuminating in that they point to how both visual and sonic affordance are perceived and experienced by the musician. First, we can relate this to familiarity and the 'intuitive' design outlined in the previous chapter. This new piece of technology is negotiated by placing his initial experience of it within his existing tacit experience of the functions other software (in this case its similarities and differences with Ableton and Logic): in other words, a familiarity with previous 'scripts' or affordance structures. Secondly, there is a mapping of conceptual classification across technologies. The reassignment of models (such as scenes and patterns) engendered by the design of the GUIs of DAWs across differing programs is illustrative of how these essentially metaphorical classifications become translated into transferable musical concepts. These visual metaphors become ways of thinking about sound, musical structure, arrangement and so on. Thirdly, the utilization of these concepts into functional tasks is understood within the wider modalities of human–computer interaction, and cash in on the metaphorical conventions of windows-based design.

Both Maschine and Ableton Live work through a drag and drop system where particular sets of sounds can be overlaid onto various types of structural organization. The idea of mapping sounds onto rhythmic patterns in this example becomes a way of ordering and approaching creativity and foregrounds the idea of dragging and dropping as a central part of the compositional process. These function/action/visualization relations echo comparable relations in a variety of other software that permeate our everyday lives in the most banal and pragmatic ways. They are similar to, say, changing font in a word processor, a visual filter in Photoshop or changing a value within a spreadsheet program, underpinned by a database-style ordering and an understanding of file organization.

Thus, Sholar's comments are part of a conflation between working with sounds, seeing and computer-based metaphor that has become embedded within musical practice. For musicians working with DAWs on an everyday basis, the ability to 'see and work' are intimately connected and imbricated. The visual facilitates sonic work; it is present in all aspects of the compositional process and provides a structural and conceptual framework for understanding that work. Further, their situation within a culture of ubiquitous computing means that the visual affordances of DAWs provide a particular type of conceptual and perceptive framing which is always present. As Marrington's (2011) study of UK music students suggests, DAWs' GUIs commonly work as a mediator of the musical idea and serve to problematize and reconfigure traditional notions of musical literacy.

---

[4]http://www.youtube.com/watch?v=xM6DaYyshyo (accessed 20 May 2011).

Visual affordance/conceptual and perceptive framing

The preceding chapter outlined how the interface design of DAWs developed over a three-decade period leading to a particular set of visual conventions. Their evolution in tandem with other types of software has meant that DAWs have always been part of a cross-functional design logic within computer culture whereby the spatial arrangement and praxis of a particular piece of software have an effect upon applications designed for unrelated tasks. Hence, their place within wider computing culture and conventions means that the visual affordances of DAWs provide a conceptual and perceptive framing which is always present within the creative process for digital producer/musicians.

Thrift and French's (2002) discussion of spreadsheets is enlightening in this regard. They examine the ways in which the development of a particular type of thinking engendered by certain types of software can lead to 'new coalitions, new forces, new realities'. They observe that the underlying premise of the spreadsheet (calculating values through the interconnection of regular formulas) provides a 'design foothold', a fundamental way of ordering from which other types of use develop exponentially from the use value of the original software. This is a pattern we can trace across differing types of computer application: for example, the database's collection of metadata which can then be recalled, creating new combinations of information is a fundamental principle of social networking.

Similarly, the functionality of contemporary DAWs draws upon a variety of design templates in terms of how sound is organized, and database organization and recall is also a key part of DAW design. While their file browsers tend to draw upon the language of conventional instrumentation and sound processing in order to label MIDI presets and sample libraries, they also create their organizing principles and utilize metadata in a ways drawn from other types of computer application. For instance, Logic Pro's sample browser (where users can access a large number of bundled samples) works through a grid that can be cross-referenced visually in a large number of differing permutations. Each section of the grid is labelled with particular criteria that rely upon an understanding of genre (world, urban, jazz, etc.), instrumentation, production conventions (dry, distorted, processed, etc.) (see Figure 3.1), intrinsic musical qualities (grooving, arrhythmic, etc.) and the affective/semiotic qualities of particular musemes (dark, cheerful, relaxed). The user may then highlight any number of differing criteria to narrow the search for a suitable sound; the more the elements of the grid are selected, the smaller is the selection of actual files displayed. The act of choosing files or sounds then is simultaneously based in an inferred understanding of musical semiotics and a learnt understanding of how metadata and database applications work. Database-style ordering constitutes just one small

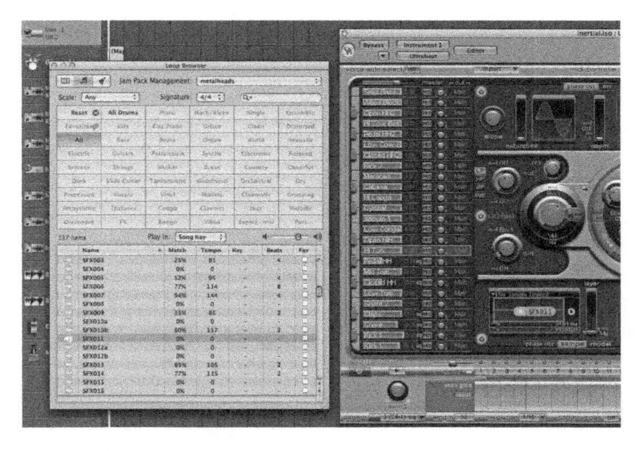

FIGURE 3.1 *Screenshot of database organization within a Logic session.*

example and the overall visual conventions of the window's interactivity can be seen to effect creativity at a more fundamental level. As Théberge (1997, 199) notes, the emergence of entirely electronic instrumentation has meant that the direct relationship between sound source and physical gesture that is central to traditional musical instrumentation has been severed, leaving the relationship between interface and sound as essentially arbitrary.

The DAW's place within a cross-functional design logic has served to replace arbitrariness with a new visual orthodoxy, and crucially, the conventions of a wider visual culture within personal computing have an effect upon how audio process is perceived visually by musicians. These conventions construct a set of spatial norms that have an effect upon what a DAW looks like, but also lead to distinct ways of working. We can broadly identify a number of strains within these new modalities relating to specific actions such as layering, freezing, drawing, cutting and pasting, looping, zooming and undoing.[5] Within the context of DAWs all of these actionable properties have become central to how sound is manipulated, ordered and conceptualized.

Layering

One key effect of the visual language of DAWs is that of layering. Again, we can relate this to the canonic affordances of the GUI outlined in the previous chapter. Windows have a pragmatic three-dimensionality; they offer depth

[5]Undoing provides a key area in which DAWs have changed workflow within the creative process. The undo function allows producers to quickly use trial/error approaches, without (of course) threatening the whole process. In some ways, this function could be seen as more 'new' than some others which are already strongly rooted in earlier (pre-digital) practices.

and to a certain extent a hierarchical ordering of information. In DAWs such as Cubase and Logic, for example, the screen offers clickable sections to reveal windows inside windows each containing differing graphical representations of musical structure, sound, volume and expression. For example, the main window within Logic is the 'arrange view', which is made up of blocks representing the overall structure of a piece organized horizontally according to the VST instrument or sound source used to create them. This window contains smaller units of graphical information such as 'midi regions', waveforms representing recorded or sampled audio and VST plug-ins. In turn, each of these elements can be double-clicked in order to reveal new layers of information: where MIDI notes are placed in a scale, the parameters of a soft synth, the shape of a waveform, etc. (see Figure 3.2). Each of these parameters can then be changed within the interface of the new window through a process of dragging, highlighting or changing values.

Such a layering of interface information is something that occurs across DAWs and is also a part of object-orientated programming tools such as Max MSP and Reactor. Both of these types of program visually encourage the user to think about the synthesis or audio processing that lies 'beneath' an individual objects' position within a processing chain. This particular form of layering gives rise to distinct ways of thinking about sound. It encourages the musician/producer to view individual blocks of sound from the outside inwards, to split up sonic processes into ever decreasing units, with distinct types of information contained within. This 'inward' perspective re-enforces the predominant perception of music as a primarily timbral practice that has been a feature of the aesthetics of differing types of electronic music. Thus,

FIGURE 3.2 *Screenshot of Logic session showing arrange view with two pop-out windows (midi region and ES2 virtual synthesizer).*

thinking about music as a timbral expression (Hugill 2008) or in terms of materiality (Demers 2010) is present in the very spatiality and affordance of the tools used in its creation.

Frozen in time

These fundamental building blocks are inevitably worked into larger musical structures. They might be used in terms of rhythmic, melodic or even atonal passages that progress into larger structured wholes in the same way that any form of musical composition takes place. However, the visual conventions of the GUI again have critical implications upon how music is conceptualized here. The normative left–right orientation of the computer screen leads to an understanding of time unfolding across the screen. While this is not dissimilar to many forms of musical notation, further conventions of visual representation of sound within the computer screen make this conceptualization distinct. Waveform displays are usually representations of amplitude rather than pitch frequency, while the individual notes contained in MIDI information are oblique, unless the user zooms in on the information. Such a visual representation of sound as frozen in time leads to a conceptualization of musical material into distinct, temporally located blocks of musical information rather than unfolding in terms of melody and harmony as it would in, say, Western notation.

The idea that composition often takes place by producers moving around blocks on a screen, and that producers 'think in blocks' has become so prevalent within electronic music culture that it is almost at the level of cliché. The title of the Canadian electro producer deadmau5's 2010 album *4 × 4=12*, for instance, is a deliberate and ironic reference to the way in which DAWs actively construct and reveal the generic conventions of electronic dance music.

> All the DAWs, the one thing that they all have in common is that you can see everything. The *4 × 4=12* thing is kind of a piss take on that because it's really kind of formulaic. You look at a 4-bar-long clip stacked to another one, and then with an 8-bar or 16-bar loop or breakdown – but never a 6-bar one. I wish I could make some kind of algorithm, if I was a talented coder, to analyze these patterns because it looks like there are so many common elements other than the sounds and melodies. But the structure of more or less every song ever played by a DJ is so damn close.[6]

While it is debatable whether the visual conventions of DAWs can be seen as fundamental in shaping the entire structural conventions of electronic

[6]http://digi10ve.com/2011/01/06/deadmau5-4x412-production-interview/ (accessed 20 May 2011).

dance music (these are after all a set of subgenres which have a powerful set of structural rules that stretch back to the 1970s; see, for example, Gilbert and Pearson 1999), there is no doubt that the DAW's visualization of structure does provide an effective monitoring of temporal progress and whether a track 'makes sense' in terms of its ordering.

The left to right visual organization of time and the vertical arrangement of tracks also encourages a collage approach to composition, where ideas are tried, monitored and set in terms of the eye. This is especially true of during editing and arrangement. For example, the process of deciding where to place a particular section within a track is necessarily articulated by means of a visual assessment and is only reviewed by ear once a particular edit has been made. Similarly, the dynamics of a track as it unfolds throughout time are also often guided and sculpted visually. For example, the ability to draw controller automation data (such as volume, panning, etc.) directly onto a sound wave or MIDI track in an arrangement window requires a visual perception of time, volume, etc. Of course, the producer is always thinking in terms of sound but there is a kind of sensory elision whereby the familiarity with a particular application means that the tactile, the visual and the aural are conceptualized and experienced simultaneously.

Looping and layering

The collage approach can also be found in the common practice of looping and layering which is often a starting point of creative activity. Sometimes this process can be of constructing dense loops through an amalgam of MIDI and sampled material layered on top of each other before a gradual process of stripping back layers each of which will form linear sections of the finished track. This is often done through a mixture of trial and error and a tacit understanding of how those stripped back elements fit into larger musical structures. These initial loops may only be two or four bars in length but they contain all the sonic elements that make up the whole or may contain the ultimate crescendo or main hook of the overall track. The track is then stripped back down to its rhythmic core and gradually built up through a series of embellishments made up from the original 'dense' loop.

The GUI of the DAW provides a naturalized visual platform to undertake this process of looping, layering and then stripping back. For instance, the successful UK techno and house producer Jamie Jones describes his working process in the following way:

> Yeah, I generally work quite fast, and I have a formula where I start off with – if I'm making a house track or a techno track – I start off with a kick drum. I usually start off with a bass sound, try and find a hook, vocal snippets, whatever, and I try and build a track, the loop up until it'll be at

> its maximum point, where the main melody or the main hook ... pretty much 90% of the elements are there and I'll just break it down to the beginning and build it back to that point. ... I usually take a bit out of the loop and put it in the beginning. I start off with a sound or a drum and then I build it up. I don't really copy the whole loop across and take things out – I take things out of the original loop and add to them. I guess the only routine I have is I'll work and work up to that point.[7]

Clearly, this mixture of trial and error is undertaken with a tacit understanding of how those stripped back elements fit into larger musical structures. Jones's comments also illustrate how these initial processes are always mapped back upon the expectations of a particular field of production (Bourdieu 1993). This technique is especially prevalent in electronic dance music where such short phases constitute the regular building blocks of a track (see Tagg 1994, 213). Hawkins (2004) identifies 'hypermetric units' (four bar phrases of 16 beats) which form part of a dynamic linearity characterized by builds and breakdowns.[8] These two elements form the fundamental core of dance music's formal and generic rules. The tension between repetition and subtle embellishment within these rhythmic units often constitute the core of a track in genres such as techno, house and minimal.[9] Such recordings often start with the very basics of its rhythmic components such as a 4/4 kick, while elements such as high-hats, snares, keyboard pads, etc., are gradually introduced. There is obviously a clear pragmatic reason for these conventions in that simple kick drums at the start of the track give DJs the chance to beat match and segue into other tracks that form the constituent elements of a DJ set. However, the gradual introduction of sonic elements constitutes much more than this. Rather, they are the core way in which a track achieves a sense of trajectory, progression and dynamic expression.

Dragging and dropping

This process of looping and layering is again facilitated by a cross-functional design convention emerging from the DAW's historical development, that

---

[7]http://www.residentadvisor.net/feature.aspx?1200 (accessed 20 May 2011).

[8]The term 'hypermeter' was coined in 1968 by musicologist Edward Cone. Cone's (1968) work points to the fact that in various contexts music has been approached in terms of blocks well before the ubiquity of DAWs, especially in oral forms of musical practices. The significant issue here is that DAWs are making this more apparent and putting emphasis on this fundamental structural vision of music within a much broader context of songwriting and production.

[9]Again, this type of layering is far from novel; for example, it forms the very basis of most oral practices in folk music and is also the case in ancient 'art' music forms, such as the passacaglia/chaconne (which starts with a single repetitive phrase on top of which more and more elements are gradually added).

is, dragging and dropping. This process often becomes something akin to constructing a sonic palette from which the final work will be formed. Again, working with an initial visual organization can help to clearly organize sound elements that can then be dragged and dropped in a constant process of moving visually, reviewing and re-editing. For example, in a video discussion of his production process, the house producer Alex Metric describes how he consistently switches between the aural and the visual in pursuit of finding the right sound.

> What I'll do, once I've found that sample and it sounds like it might vaguely fit or there's just something about it I like, I'll bring it into Ableton get it timed up, get the whole track timed up and then just start like pulling chunks out of that track. I like to get a few different possible loops and little bits to use. … And I'll start just laying them next to each other, and then I have like a little collection of possible loops that are gonna work and I'll drop them over the original track. … I always find that random is the way that you get the best results, so I just move things around with no sort of thinking behind it and that's how you end up getting the best little bits. So I'll get if you start moving things around, randomly then the right loops appear.

This quote is illuminating in a number of ways. Again, it clearly demonstrates an almost instantaneous modulation between visual and sonic affordance: what sounds 'fit' with an idea of an overall creative project are organized visually and then can then be dragged and dropped in a constant process of moving, aurally reviewing and re-editing. The quote also illustrates how aspects of mental play central to the creative process become bound up with the functionality and terminology of the particular technology used. The pragmatic elements of experimentation are expressed in the metaphorical language of the virtual and through its visual realization. Terminology relating to spatiality ('moving things around'), units of musical and sonic material ('loops' and 'chunks'), individual aspects of instrumental arrangement ('layering') and overall musical structure ('tracks') are less translations of musical concepts as synchronic conceptualizations of process in themselves.[10]

Copy, cut and paste

This constant process of moving visually, reviewing and re-editing as part of the creative process is facilitated by another central cross-functional design

[10] Metric's comments also illustrate how the core functionality of an individual program can facilitate particular ways of experimenting. Ableton Live's automated rhythmic analysis of samples as they are dragged and dropped into its 'session view' means that they become tied into a session's global tempo without the need to manually stretch it into time.

element: cut and paste. While cut and paste might have been a conceptual (and pragmatic) element of a number of differing strains in popular and avant-garde musics, its central (visual and functional) position within contemporary DAWs shifts its focus somewhat. Cut and paste becomes one of the fundamental organizing principles of how a musician engages with the interface of a DAW and thus foregrounds its importance within musical practice. The reduction of sound into easily movable graphic blocks helps to facilitate trial and error as a central part of the creative process but also serves to encourage users to think visually about a project. The ease in which differing elements can be tried, reviewed, resituated and deleted at the click of a mouse clearly speeds up workflow, but again imposes particular types of spatiality in the way sound is conceptualized.

With DAW technology, cut and paste actions become more central and visual but they also become more precise. On the one hand, this has a pragmatic advantage, in that visually zooming in affords new levels of control by giving a larger visual surface area and ability to achieve a magnification of the sound file, which allows for an accuracy unavailable in equivalent hardware. For example, UK techno producer Perc (real name Ali Wells) talks about the changeover to a computer-based recording and sequencing studio in terms of precision:

> Technology has helped me a lot, the switch from a fully hardware based studio to Ableton interfacing with a few choice pieces of kit gives me a flexibility that I could not have imagined before. For remixing the ability to creatively and accurately edit audio visually has been a massive change for me, so much better than staring at the screens of samplers and grooveboxes.[11]

DAWs then engender new levels of accuracy in the sampling process, but, perhaps more significantly, facilitate a shift in the actual types of sound that can be produced. For example, there is a clear connection between the ability to see minute sections of sonic information on screen and the development of distinct aesthetic trajectories within electronic music. Zooming in is another related cross-functional design element appropriated from visual software that has been fundamental in shaping technique. Zooming in on a particular region of an audio file allows for a degree of focus that would be extremely difficult with other hardware-based ways of editing samples. This ability has led to the emergence of whole subgenres such as microhouse and glitch which are facilitated through microsampling of minute sections of audio information and using the resultant sonic materials as a basis of their structural and aesthetic conventions (Harkins 2010; Monroe 2003).

---

[11]Perc interview from Earwiggle Dublin, http://technomusicnews.com/perc-interview-from-earwiggle-dublin/

The development of these new sounds and the development of a broader digital aesthetic within electronic music and production form the basis of the final chapter.

Just as DAWs facilitate cut and paste as a central part of the composition progress process, so they engender a number of ancillary creative actions that have become central to the emergence of sonic characteristics of electronic music produced within these contexts. Stretching clipping, looping, cutting and zooming are all actions facilitated by bodily and visual purpose that are resultant in a particular set of sonic characteristics that become generic markers.

Summary

This chapter has suggested that sonic and visual affordances are perceived together in the moment of creativity and that musicians and producers are engaged in a highly developed level of conceptual integration. In other words, the coalescence of visual, interactive and musical comprehension provides a kind of perceptual blending through which the solutions to given problems are realized. The use of accumulated specialist musical knowledge is necessarily inflected with an embedded way of structuring knowledge per se, formed through a prolonged and everyday interaction with computer technologies. The producer/musician is always thinking in terms of sound as a final outcome, but as the dual application of the ecological approach outlined here has illustrated, there is a kind of sensory elision whereby the familiarity with a particular application means that the tactile, the visual and the aural are conceptualized and experienced simultaneously. The actual use of sound in this matrix is developed in the following two chapters when the theoretical framework relating to musical and sonic affordance is applied to the social/institutional situation of the cultural producer and the aesthetics of contemporary pop/electronic music respectively.

Hence, for musicians and producers, DAWs are not just a way of processing or organizing the product of creative action. They are in themselves, actors within the process of creativity, helping to shape and direct particular ways of working, thinking and doing. The legacy of the particular historical trajectory outlined here (both in terms of the general design logic of GUIs and the specific development of computer-based DAWs) is always present within the creative act. As Chapters 2 and 3 have demonstrated, the specificities of the historical context in which DAWs evolved has meant that the embedded metaphors and generic visual architecture of computing have been constant throughout their evolution. This has meant that rather than a repositioning of working practices related to analogue technologies which musicians and technicians would have worked with in a pre-computer-based studio, the DAW has been resultant in a new set of modalities.

Ultimately, what I want to suggest is that musical creativity has to be understood in relation to a wider context in which the social, the technological and the musical are in a constant state of overlapping. The new ways of conceptualizing and working outlined here are also indicative of wider perceptual shifts occurring across differing aspects of culture related to the changing nature of HCI. They employ conceptual, visual and tactile modalities which, through our ubiquitous use of computers, punctuate our quotidian moments of work and leisure, of engagement and distraction. For musicians and producers working with DAWs, then, creative practice has become part of a wider interactive engagement that can permeate a large portion of our waking lives.

CHAPTER FOUR

Creativity as discourse/creativity as experience in electronic dance music and electronica

Field Recording for Electric Blanket, *Liverpool March 2011*

I am crouching on the pavement of one of the quieter sections of the Dock Road, Liverpool's five-mile northerly scar to the trauma of its industrial heritage. Although there are many remaining hubs of activity in intermittent warehouse and dock buildings, the area is sparsely populated. The occasional echoing clang of a moving crate or conveyer belt punctuates the bird noise and constant traffic hum. I am here trying to capture some audio trace of the city's past, some intangible sonic patina that I hope will transmit something of the sense of change, loss and nostalgia that I have felt from some of the people I have interviewed over the past six months.

The field recording I am undertaking is for an audiovisual (AV) installation that will be exhibited at FACT (the Foundation for Art and Creative Technology) in Liverpool later in the year. My friend and long-term artistic collaborator Bob Wass and I have been commissioned to work with residents of sheltered housing in the city to create an AV piece around the theme of sound, memory and place. After interviewing twenty residents about how sound might be instrumental in evoking memory we are now in the process of capturing field recordings and film of some of the places they identified as important or remembered as the location of significant sound events in their lives. Of course the vibrant soundscapes they recalled are now long gone, a fact that the sparseness of the soundscape in my headphones only serves to reinforce.

Listening becomes very intense, my ear is attuned to every aspect of the soundscape in concentrated detail. The concentration afforded by the project makes the sounds more intense; I 'hear' elements of the soundscape that my brain would otherwise screen out. I am also listening in a different way; in terms of utility and aesthetics. In this context, I evaluate each element of the soundscape in a matrix of possibility. How will it fit with the film footage we have already shot? Is it interesting in itself? How will it fit with other sounds I have already recorded? Does it have particular tonal or rhythmic qualities that will cut through the general ambient hum of the soundscape?

At this nascent stage in the process there is also an assessment of what those sounds will signify for a gallery audience, how they will work in terms of an audiovisual installation, how they will sound in a gallery context and, perhaps subconsciously, how the eventual soundscape will chime with the aesthetics of the 'genre' that I am working within (in this case contemporary digital art).

Beginning a chapter which primarily deals with creativity in the context of electronic music production with a reflection on the making of an audiovisual installation may appear counterintuitive. However, the way that the specificities of this particular project caused me to focus upon the sounds of the external world as possible elements within a creative end product is indicative of the issues that I want to open up within this chapter. What I heard on that early spring afternoon was consistently mediated through my own situation within the sedimented knowledges and modalities that have already been outlined within this book. That is, I was constantly aware of what could be done to those recorded sounds once I had taken them back to my project studio. My listening was also subject to another layer of mediation: a tacit understanding of the aesthetics and conventions of the 'genre' I was working within and an expectation of the spatial and experiential feel of the space in which the work would eventually be exhibited. During the process of sound capture, I was thinking about how the end product would be experienced by an audience. In this case, the 'genre' was audiovisual digital art and the space was a specially built black room that we had discussed in advance with the production and technical staff at the gallery. On another level, my assessment of the soundscape was framed by my own internalization of what it means to be creative. In other words, how could those particular sonic elements come together in a meaningful and original way that I would find creatively satisfying.

The elements that informed my listening (and recording) that day could be loosely categorized under four main headings: technological possibility, an understanding of the field (in terms of institution, genre, spatiality and audience expectation), sonic affordance and wider discourses of creativity. All of these elements provided a distinct framework of experience in that moment, an experience marked off from everyday life through an aestheticization of the world around me and a specialized mode of attention. With these four elements in mind, this chapter aims to examine notions of creativity in relation to digital technologies and ask how creativity is experienced by the digital musician/producer. In order to unpack the specificities of the implications of social context upon creativity it focuses upon producers working in electronica and electronic dance music. This particular generic framework allows for an examination of the nuances in social context that influence creative activity by drawing upon very specific case studies and examples. The chapter first examines creativity as an experience and a socially constructed discourse before going on to suggest how both are played out in relation to genre, technology and spatio-social context.

Throughout the history of popular music, the concept of creativity has provided a notoriously knotty set of associations. Even today, common-sense notions of creativity continue to have currency within Western popular music culture. McKintyre (2008) succinctly conceptualizes these into two main modalities: the inspirational view based around the idea that the creative individual must reach or induce an illogical or irrational state in order to induce inspiration, and the Romantic view based around ideas of the 'extraordinary and the use of innate gifts of intuitive talent ... of the unconstrained, self-expressive, quasi-neurotic artist existing in their garret, waiting for the muse to arrive or for inspiration to strike' (2008, 41).

Critical scholarship on popular music and creativity (Toynbee 2000; McKintyre 2008; McKintyre and Paton 2008) has tried to move away from romanticized and inspirational notions of creativity towards a more grounded theoretical framework that recognizes the situation of the artist within a distinct social, artistic and historical context. Within this theorization, creativity takes place within particular social formations that have an active effect upon creative choices and ultimately the end product of a particular creative endeavour. Cultural practitioners accumulate a store of knowledge in order to be fluent in the skills, conventions and histories relating to the cultural practices in which they are engaged. These knowledges are then put into action during the creative process.

Nevertheless, the voices of artists still consistently accord with the inspirational view, in that often they represent their own practice as being punctuated by fleeting, almost inexplicable, moments of inspiration where the process 'just happens' (see Zollo 1997; Tucker 2003). The everyday persistence of the inspirational and Romantic views of creativity is perhaps bound up in the power and the apparently intuitive nature of the creative process as an experience. For example, Zollo's (1997) collection of interview material with songwriters is littered with accounts which present the songwriting process as a quasi-mystical experience where a creative individual is seen to be channelling or merely witnessing creativity from the ether (the inspirational view of creativity as a form of self-representation). It is perhaps easy to dismiss such muddied thinking about creativity as irrational or deliberately obfuscating the process in order to uphold a central romantic ideology within popular music culture. Frith (2011) concludes that the concept of creativity itself is so bound up with such Romantic discourse that it is ultimately a hindrance rather than an aid in understanding music making practice and argues that it should be abandoned entirely within sociological enquiry. However, to discard these moments entirely (and the idea of creativity more broadly) is to overlook the way in which the act of making music is *experienced* as being creative and the power of the idea of creativity in informing and guiding what music makers do within their practice. What I want to suggest here is that discourses around creativity are central in setting up a series of tensions that impact upon the choices these individuals make, how they view differing aspects of their work, and ultimately how they experience making music.

Creativity as experience and discourse

An approach that recognizes the social production of creativity while acknowledging that the act of making new music involves certain types of intense experience is not inherently contradictory. In fact, such intuitive moments of creativity are innately bound up with the social, in that they involve the internalization (and naturalization) of a complex set of modalities, codes and discourses that are experienced as a simultaneous composite. Such an understanding of creativity as experience has echoes of the work of the formative nineteenth-century pragmatist philosophy of John Dewey. Dewey (1934) posited experience as relational: as involving a reflexive and active connection between self and the external object. Experience, therefore, involves the engagement of the individual (in terms of the sensorial, emotional and cerebral) with the environment that exists outside her subjectivity.

Such an interplay with the environment is, I think, crucial in rethinking musical creativity in terms of experience. Most obviously, music making takes place not only through the mental processes of the individual, but also through the use of particular technologies and in the reconfiguration of sounds and musical structures that have a physical presence within the external world. In other words, the cognitive processes involved in creating music are as equally bound up in the perceptual as they are in particular ways of thinking.

What I am suggesting, therefore, is a relational approach to creativity in which the experience of being creative is understood as an interplay between the individual's situation within a particular field of production and the affordances of sound and technology. Hence, my use of the term 'creativity' here not only refers to what cultural producers do in the act of creating, but also takes into account the discursive traditions that construct ideas of creativity. The aim is not to demarcate what is creative or otherwise, but rather to explore how creativity as a discourse has a fundamental impact upon practice and how the process is experienced. As Frith (2011) argues, the notion of creativity itself is a culturally relative construct. He notes that the idea of creativity is a distinct product of 'societies in which there is a particular sense of selfhood and the valorization of the new. Creative freedom is not something that people naturally aspire to, as part of their humanity' (Frith 2011, 70). His ultimate conclusion is that within 'capitalist societies musicians are *constrained* to be creative, both culturally and as a matter of political economy' (ibid.).

So, on the one hand, creativity refers to a set of practices that have economic and social value according to institutions and a wider market economy, and on the other, it is a culturally specific and constructed idea about what it means to be creative. These dual definitions of creativity as differing types of 'social fact' (i.e. socially constructed discourses and norms which exercise constraint) are significant. They are social facts that cultural producers simultaneously carry with them and are fundamental in structuring creativity as an experience. They shape the experience of creativity by setting up a series of tensions: tension between the ideology of innovation and the demands of particular institutions, between the structural constraints of genre and the drive towards originality, between the imagination and the limitations of a particular music technology. These tensions inform the self-definition of the individual *as* creative and what they think they are doing by *being* creative. In other words, the space between these points provides a stage for creativity to be played out; the dual discourses of innovation and formal/institutional convention not only serve to place constraints on a given creative outcome, but also act as a spur that the cultural producer works against. Creative tensions, then, provide a filter through which sonic affordances are put into action.

The situation of the creative work within an institutional context has long been a concern within the sociology of the arts. Studies across differing artistic fields have highlighted how 'art worlds' (Becker 1982) sponsor and validate artistic works and ultimately shape their final fruition (Zolberg 1990; Du Gay 1997; Hesmondhalgh 2007). The intersection of field and individual has, in turn, been a central issue within studies of creativity. Csikszentmihalyi (1988) formulates these issues into the domain (the knowledge system a creative individual uses), the field (the cultural spaces, peer group and institutions centred around the domain) and the individual agents who operate within these cultural spaces. A similar approach to creativity has been taken up by further studies which attempt to examine what musicians do within the creative process (Toynbee 2000; McIntyre 2008). In his study of creativity and popular music genres, Toynbee (2000) proposes that the musician works with certain socially constructed codes and practices in order to be 'creative'. Borrowing the term 'space of possibles' from Bourdieu (1993, 64), Toynbee proposes the concept of 'creative space' to describe the palette from which the musician works. First, this relates to the artist's habitus of socially constructed subjectivities. Secondly, it relates to the historical fund of practices, the textual forms and codes which Toynbee terms the field of works. Creative space is a combination of these two things and the relationship between them. Toynbee conceptualizes this as a 'creative circle', which he terms as a synchronic model for trying to explain creative action at one particular time. Similarly, in his work on creativity among popular music songwriters, McIntyre takes Csikszentmihalyi's (1988) systems model approach. For McIntyre, these individuals are at the heart of artistic progression, in that through making particular creative choices they are active in creating 'variations in the symbol system' (2008, 49), although they work within constrictions of the domain with the result that generic or creative change tends to be characterized by a series of small increments.

It is important to recognize that such ideas about creativity are always culturally constructed and relative. What particularly interests me here is the position of innovation and variation as socially constructed discourses that work their way into the subjectivities of cultural practitioners. In one respect, creativity as a discourse constructs a tension between field and the self-definition of the cultural producer as an autonomous artist. Artists often see themselves as working between the needs and constraints of the field (or the parameters of a given project) and their own compulsion to 'be creative' in terms of progression, innovation and newness.

Creative hierarchies

By way of illustration, I want to briefly turn to two examples where electronic musicians have collaborated on soundtracking projects. Such examples are useful as the constrictions of working towards a mixed media text serve to clearly highlight tensions between an individual's idea of their own creativity, necessities of form and the demands of particular fields of production. This became apparent during a panel discussion which took place at the Unsound festival in Krakow, Poland in 2010. The panel featured Lustmord (a key figure in the dark ambient genre that emerged from the industrial music scene, who has also worked as a sound designer for big-budget Hollywood horror films), Ben Frost (Australian electronica producer who has undertaken various soundtrack projects), Polish industrial producer Zenial and Raz Messinai (electronic music producer and soundtrack artist). What started out as a discussion of the festival's overall theme of horror quickly shifted into a broader discussion of motivation and inspiration. As they swapped anecdotes, the panellists began to collectively move towards a fairly negative appraisal of working within certain institutional contexts. All four had experience of film and video game work and the tone of the conversation involved a discussion of how restrictive they had found composing for these types of media. They then went into a detailed discussion of how to negotiate and work within the expectations of these third-party institutions that were paying for their creative labour. In each case, the artists articulated their position in regard to the formal conventions of film and video and their conception of themselves as creative individuals. Interestingly, despite being set in opposition to what they saw as their 'real work' (i.e. producing various types of experimental electronic music) these constrictions came across as both restricting and a spur for creativity.

Frost, for example, described working on a video game project. He indicated that he was both working with and against the constrictions and conventions that the medium imposed in order to facilitate his own creative process. On the one hand, he acknowledged that there was a clear creative 'challenge in creating music which doesn't have a clear linear narrative as that's a major part of my work'. Music for video games has to be composed in a cyclical way as a player spends an indefinite amount of time within a particular environment. Thus, the particular outcome required for the project posed a particular set of problems. This made him work in a new way, responding to the challenges of a new set of restrictions. On the other hand, he talked of deliberately working against the sonic conventions of the genre by 'sneaking sounds in there that don't belong, incongruous with what you're supposed to be doing' in order to make the process more interesting for himself and to (both sonically and discursively) align the project with the rest of his work. These constrictions, as well as the very prescribed and formalized notions of what music should 'do' within the context of audiovisual media then, tend to rub up against an artist's self-definition as a creative individual and an overall understanding of their own practice.

A second example of this tension was articulated by Matthew Herbert, an internationally renowned electronic producer/musician, who has undertaken numerous film music commissions. Hebert is best known for his extensive use of field recordings (often with set thematic parameters[1]) in creating post-techno-influenced electronic music. In a conversation with the author, Herbert described how he constantly tried to integrate his overall working strategies into the soundtracking process by producing scores that manipulated a film's existing audio tracks and Foley[2] into musical pieces. He indicated his frustration that every time he had attempted to do so the results had been rejected by the filmmakers who wanted him to produce something more 'filmic' in tone. Citing one example of a film that he worked on, he commented:

> When they got in touch they were like 'Oh we're really big fans of your work' but ultimately what they wanted was very conservative ... romantic, orchestrated stuff that's become written into the language of film music. Anything else they found difficult to handle. And you end up thinking 'You know what I do, what I'm known for, so why ask me?'[3]

Both of these exchanges struck me as interesting in a number of ways. First, particular conceptions of creativity were articulated through a division between different compartmentalized aspects of their work and through discrete projects. On the one hand, their creativity was described as something that may be harnessed and put to use in a variety of ways. However, there was a clear hierarchy between projects, with their 'real' work (the production of recorded music) being prioritized over other project-based work. There are clearly pragmatic institutional issues at work within this distinction partially to do with scale. In terms of their record releases, each of the artists had developed relationships with small niche market labels where they are afforded a great deal of creative freedom and are able to direct their work according to much broader (and looser) institutional pressures. As I have argued elsewhere (Strachan 2003, 2007) the working practices of micro independent record labels provide a much more diffuse set of constrictions (allied with a significantly reduced chance of significant financial recompense) upon their artists, lending them a degree of autonomy which would be impossible within the context of larger companies. It was clear from the discussions that this aspect of their work was regarded as a self-directed and open articulation of their creativity which was intermittently captured through the formats and release conventions of the institutions with which they were working. In contrast, the application of their creative labour to the film or gaming industries involved a much clearer set of constrictions: deadlines, set budgets, layers of institutional discussion, and importantly, the formal and aesthetic paradigms of the media with which they were working.

---

[1]For example, *Around the House* (1998) which sampled household objects, *Bodily Functions* (2001) which was made from recordings of human internal organs, skin, hair, etc., and *Plat Du Jour* (2005) which used sounds from food production.
[2]Recorded ambient and action sounds which are synched with film to give a sonic realism.
[3]Personal communication with author September 2008.

In both examples, the urge to compartmentalize between differing types of project was underwritten by a defined idea of their own creativity. In one way, these hierarchies can be seen as in keeping with long-held discursive conventions relating to art and commerce which are central within certain popular music cultures (Stratton 1983; Negus 1992; Strachan 2007) and tensions which pervade the cultural industries more generally (Hesmondhalgh and Baker 2010). Here, a clear hierarchy emerges where some activities are regarded as a more appropriate conduit for their creativity than others. In another way, constructed discourses of creativity provide an overarching foundation in terms of the pragmatic choices made by individuals in the process of making music. For both Frost and Herbert, discourses of an idealized creativity were integrated into whatever project they were working on – serving to validate the experience of the process as *creative* rather than on a spectrum moving towards alienated labour. Even when faced with a differing set of constrictions and institutional pressures, the urge to be progressive or to subvert form (Frost's 'sneaking' in of sounds that 'didn't belong', Herbert's urge to push at the boundaries of film music convention) underpinned how the projects were approached and how sonic affordances were acted upon. The way in which the institutional and formal tensions were negotiated within these examples is indicative of how individuals actively construct creative subjectivities, that is, how they understand the processes in which they are engaged and how a broader creative subjectivity mediates, moderates and monitors the nuances of creative action undertaken by an individual.

Underpinning these ideas is the autonomy of the creative process as a valuable experience in and of itself whatever the project. In a later conversation with the author, Herbert positioned his own creative process at variance with the very institutions that support and validate his music:

> I think that one thing that the music industry is colossally bad at is selling the idea of music as a process rather than a product. So it's all about what Madonna's next album is, which is what it is. It's this blank thing that's all shiny and polished, there's no mistakes and it's exactly absolutely what it is. You have it like a ham sandwich and it's gone. There's never a real recognition that real music is a process … that art is a process. So you have to fight against these impulses [in your day-to-day practice].[4]

Genre and social use

The above examples provide particularly polarized examples of how discourses of creativity intersect with the institutional and aesthetic pressures of a given field. However, an insinuated split between 'real' and secondary

[4]Personal communication with author September 2008.

creative work identified by the artists on the Krakow discussion panel was relative, and does not mean that their purely musical projects were somehow an expression of free and unfettered creativity. Creative tensions and the internalization of discourses of creativity are equally discernible, and intrinsic to, all areas of music making.

One of the fundamental areas where these tensions occur is through the prism of genre. On one level genre provides a benchmark for the acceptance of a particular piece of music and a set of rules that are simultaneously taken into account and intermittently broken in the pursuit of progression and originality. Toynbee (2000), for example, points out that genre provides a 'necessary starting point for creative action' from which it is very difficult to break away. This does not mean that music is just endless repetition; rather genre is the space where the tension between repetition and difference is regulated. Within this model generic change occurs when new and less likely choices are made: choices that break through generic boundaries and move a genre forward or become codified as new genres or subgenres. In Toynbee's terms the 'social author' has creative agency in instigating generic change, in affecting the trajectories of established genres or the creation of new ones. However, as Fabbri's (1982) classic text on the subject indicates, genre is not just a matter of sound itself but is bound up with a variety of contextual frameworks, a series of semiotic, ideological, formal/technical, economic and behavioural 'rules'. This theorization is important as the discursive practices relating to creativity outlined above provide a core element of the ideological rules of many popular music genres that, in turn, impinge upon how formal and technical elements of genre are handled by the creative individual.

In order to unpack these issues I now turn to examples from various types of electronic dance music. Dance music provides a pertinent way into thinking about the particular constrictions of a field in that it has a clearly defined purpose and carries within it a powerful discursive legacy relating to creativity and progression. We can relate this historically to two concurrent strains within dance music culture: an engagement with pleasure, the body and eroticism (Fikentscher 2000; Hawkins 2004) on the one hand and a tacit avant-gardism on the other. These two strains have sometimes been seen as essentially conflicting, with a discursive split occurring in the 1990s with the emergence of various post-techno and house musics. Gilbert and Pearson (1999, 76) note that 'such unfortunately titled labels as intelligent techno' signalled the removal of these musics symbolically from the zone of the body elevating them by association with notions of intellect and "art"' and that institutions such as the music press and art galleries were instrumental in 'deploying a muscular modernist discourse on electronic music'. However, this critical separation constitutes something of a false distinction and I think it is erroneous to draw such as stark polarization between the aesthetics of electronic musics ostensibly designed for different listening experiences. Electronic listening musics (Marstall 2002) still have a clear connection with bodily affect (Strachan 2010) and experimentalism is central to the practice of many dance music producers. Sonic experimentation has been a key component within dancefloor-orientated musics whereby novelty and the introduction of new sounds have been an integral part of dance music cultures leading to multifarious generic mutations based on the idea of progression (Monroe 1999). In this sense, electronic dance music is imbued with concurrent aesthetics whereby progression and experimentation have to be balanced against social use.

Rule breaking

There is clearly a sedimented awareness of this historical legacy by dance music producers and an experimental aesthetic underpins how they make pragmatic decisions during the production process. For example, in a discussion with the author about newness and genre, Dublin-based techno producer Donnacha Costello positioned 'rule breaking' as a central part of the creative process and what it meant to be creative within dance music culture. Further, he described the progression of individual genres as a convergence between an acquired sense of history intermittently punctuated by new sonic developments (sometimes the result of naïve experiments or mistakes) which were then incorporated and adapted into practice through an acceptance within the field. For Costello, this occurred through a combination of:

> people who've being doing this for a long time and then younger guys coming in who've only being doing things for two years. And I think that these guys who are just starting if they don't know the history, they might do something totally off the wall and totally out there and that will move things forward a little bit. But then I think that it takes people also who understand the history of it [electronic music] and understand the context of it a little more to kind of be able to extract what's good from that [new] development and use that you know. So it's the old kind of notion that you have to know the rules to break them effectively. So coming into it without knowing anything you can break every rule in the book but you don't know which rules you're breaking and you don't know why. But then maybe someone else who's being doing it a little longer can look at it and say 'well that was a good rule to break' and 'It was broken really well and maybe I'm going to follow that avenue'. And so you have these two streams of people both influencing the development and in turn the history of music.[5]

Two related things emerge from this quote in terms of the relationship between individual and field. The first is related to how discourses of

---

[5]Personal communication with author October 2008.

creativity feed into the practices of musician/producers and the second is how the results of that practice are validated and integrated back into the field of production. The awareness of social uses within creative practice clearly provides a tension between familiarity, difference, innovation and acceptance within their respective field. In short, all artists with aspirations towards originality or experimentation have to traverse a line between novelty and recognition. At the heart of this tension is the (perceived) listener. As Bourdieu notes, even the most avant-garde of artists recognize that their work must resonate with an audience even if this must be an 'alter ego' or ideal listener who is able to recognize the 'autonomy of the creative intention' (1971, 165). Because of its historical and discursive legacy, electronic dance music provides such a space where the autonomy of creative intention is highly valued and provides an underpinning of what it means to be creative. In other words, the breaking of rules provides both a discursive ideal and a pragmatic creative strategy. Thus, a key creative approach for many producer/musicians is to consciously attempt to go beyond the expected musical and sonic limitations of genre through a process of trial and error. This is an approach that has an active effect upon the affordance structures implicit to the creative process within electronic music. Here, the creation and assessment of sound, choices about what to retain and develop, about what to reject and what has possibility are made through the prism of experimentalism and progression of existing form.

Trial and error and experimentation are a key part of the creative process for popular musicians across differing genres (see, for example, Leonard 2010). Experimentation is an aspect of mental and musical play which musicians tacitly learn though their interactions with technology, sound and musical convention that are brought together in personalized working practices through which creativity is experienced and rationalized. However, it is important to note that this is not a random process where *any* choice can result in a new innovation within a final work. As Costello's comments indicate, there is a constant process of mapping the results of experimentation back onto musical and social frameworks. Experiments are necessarily mediated by the domain (i.e. the store of historical information the producer has) and the field (i.e. how it is accepted within the specific culture in which it exists) and the worth of new developments are always judged against what has gone before. In other words, the results of experimentation or sonic innovations in an eventual musical work make little sense if they bear no relationship to a tradition.

Hence, while the concepts of tradition, experimentalism and innovation may seem to sit uneasily together, there is a sense in which they are closely interwoven and can be identified as central to wider culturally constructed discourses of creativity. For example, Csikszentmihalyi (1988, 326) sees variation and novelty as a crucial component of creativity to the extent that it is 'impossible to judge creativity without an understanding of historical context'. Similarly, Negus and Pickering (2004, 101–2) argue that tradition is, in fact, central to the idea of originality. They attempt to problematize the perceived split between modernist and traditionalist positions on art and culture by arguing that the individual can only be original in relationship to some existing tradition and that, contrary to pervading discourses of modernism, an enduring value of tradition 'lies in providing opportunities for its extension and transformation' (ibid.).

Creative subjectivity

Genre in this context provides a marker against which producers simultaneously judge their own work and pull against in order to justify their work in terms of discourses of creativity. In this sense, ingrained discourses of creativity form part of the habitus (the embedded dispositions and ways of acting) which nuance the trajectory of an individual's intersection with the field within the radius of creativity (Toynbee 2000). To illustrate this I will now turn to the example of the personal creative history of the Berlin-based British dance music producer Sam Shackleton. The position of his creative practice with regard to the cultural field of dance music in the 2000s belies a complex relationship with genre. In one sense, the convergence of his overall production sound with emergent and highly popular genres gave him a platform to be heard and acted as a fillip to his musical career. He released his first production in 2004, the same time as dubstep was emerging as a distinct dance culture in the UK. His tracks released through Skull Disco (an imprint which he co-owned with the UK producer Appleblim) initially found an audience among dubstep DJs while also chiming with the minimal techno sound that became ubiquitous across house and techno dancefloors all over Europe in mid-2000s. Ricardo Villalobos' 15-minute remix of Shackleton's 'Blood on my Hands' became a pan-European dancefloor hit in 2007 and he went on to release material for the established techno label Perlon in 2009 and a successful *Fabric Live* release in 2010 (a CD series affiliated to the London club of the same name which became one of the most successful brands in techno music until its closure in 2016). Success within these scenes allowed him to make a full-time living out of music by plugging him into the European club and festival circuit.

In another sense, the idea of genre (and being generic within his productions) was something that provided a negative criterion through which Shackleton monitors his own work. In an interview with the author, he specifically positioned his creative process in direct opposition to an overly generic approach to production, while acknowledging that his work was necessarily situated within the social uses of dance music.

> I don't try and make music of a genre and I think once you have that approach you put limits on yourself. I'm not saying that I'm going to make a completely abstract album of Mongolian noseflute music ... I know I'm

> not at the vanguard of the avant-garde or something like that, I know that. Essentially I'm making beats music for people to dance to. But by the same token once you start limiting yourself and saying well 'this is how you make a dubstep tune' I just think for me that's the biggest turn off – I'd feel fraudulent and it's not really me.[6]

In this context the artist's position in relationship to genre is articulated in terms of the personal authenticity and creative integrity. Throughout our conversation he consistently indicated that it was necessary to position himself in opposition to the generic in order for the production process to be a fulfilling experience in itself. We can see this as an example of how discursive constructions around creativity underpin what we could term a creative subjectivity.

The feeling of authenticity within the creative act is a product of the intersection between habitus and field whereby one's actions are mediated through a developed and embodied schemata.

The habitus of the individual is subject to a highly specific and personal trajectory, constructed through a lifetime of engagement with differing musical works, differing cultural fields and discursive practices. For example, despite his rejection of genre as a creative goal, Shackleton's creative trajectory is nonetheless characterized as a series of nuanced cultural encounters with varying institutions and generic alliances. From a background in post-punk bands through to the London squat scene of the 1990s he has maintained an interest in reggae and its hybridized musical offshoots. Shackleton had already been producing music on his own before he came into contact with what would become dubstep club culture. After being in a dancehall-influenced act on the squat party circuit, he eventually made the transition from hardware production technologies to a computer-based studio environment and continued making what he described as 'bass heavy' tracks. Within a similar timeframe, he found that he felt a sonic affinity with the music being spun at FWD>> and DMZ which were influential clubs in the evolution of the dubstep scene in London.

Each of these encounters in his musical life can be seen as carrying distinct aesthetic and discursive predilections: the eclecticism and experimentalism of post-punk culture, the foregrounding of the materiality of bass frequency within reggae and dancehall and the implicit avant-gardism of electronic music. This distinct personal trajectory is indicative of how the habitus of the individual serves to construct a sense of the creative self and a synchronic aesthetic amalgam that form a key element of the affordance structures of creativity. In other words, it is through such creative subjectivities that sonic and musical affordances are put into action in distinct ways.

---

[6]Personal communication with author February 2008.

Field and sound

Implicit within the examples I have drawn upon so far is the presence of the sounding object in the manifestation of discourses of creativity. In each case, the use of sound served to articulate and verify such discourses within the institutional pressures and generic conventions that are implicit to a given field of production. What I would like to suggest here is that the key intersection of habitus and field within the creative process for electronic music producers is through an assessment and utilization of sound. The pragmatic and technological conventions of electronic music actively construct such a relationship. Working with synthesis, samples, and found and recorded sound from the very beginning of the creative process means that the musician/producer is working with material that already possesses sonic affordances. As such, particular sonic materials may 'afford' certain use values and creative trajectories that may be followed or rejected by the individual.

The sonic affordances of this material are situated within the wider culturally constructed conventions of music and the relationship of sound with the physical world and the body. For example, DeNora (2000) argues that musical material affords differing types of bodily movement or reaction in embodied action through which users experience their bodies in new ways. This might relate to dance, movement or a variety of bodily processes, but she goes on to note that this is always mediated through a process of interaction with a number of factors. She notes that 'what is key here is how the music is, or comes to be, meaningful *to the actors who engage with it*, including such matters as whether the relevant actors notice it' (2000, 49). Similarly, Clarke's use of musical affordance presupposes that the 'perceptual specification is in a reciprocal relationship between the invariants of the environment and the particular capacities of the perceiver' (2005, 44). In other words, the material and the social have a simultaneous presence in the act of perception.

It is in this coincident perception of the materiality of a given sound and its relative position in relation to the complex personal trajectories such as the one described above that individual acts of creativity take place. Further, it is through the relational complexities of the material, the discursive and the subjective, that creativity is rendered as 'an experience'. By situating creativity in such a way, I want to hold on to Dewey's (1980) central distinction between experience as a continuous unfolding interaction between environing conditions and living being, and 'an experience' which, though equally a product of interaction, has a sense of running a particular course towards fulfilment and has its own 'individualizing quality and self-sufficiency' (1980, 35). Clearly, the concentration of action, the sense of movement towards a final outcome and the element of problem solving central to action during the music- making process firmly situate creative action in such terms.

Creativity as aesthetic experience

Dewey (1980, 37) also notes that individual experiences have a certain unity that overrides the emotional, practical or intellectual constituents that make up the encounter in an overall experiential quality. I think we can further extrapolate from this position that the specificities of environment and subjectivity across differing types of music making produce particular *types* of experience with particular types of quality. They are a form of what Dewey calls aesthetic experience; that is, refined experience that is marked off from everyday or ordinary experience through particular convergences of environment and individual. Creativity as an experience for electronic musician/producers involves a coming together of sonic affordance and creative tensions resulting in a sensory and cerebral immediacy whereby such moments are encountered as an instinctive and naturalized whole. Shackleton, for example, described his working process as beginning with a meticulous construction of beats through the manipulation of samples and synthesized sound that was experienced as an intuitive assessment of the sonic:

> I listen to it, if a beat sounds good for me the tunes sort of write themselves in a way. They register with my ear and it's like 'yeah that's good'. And when something doesn't sound right it really jars with me. It might be just one little hi-hat out of place or something I'm very ... I just know what it should sound like to me.[7]

This comment also hints at two key specific characteristics of music making as a creative experience. First, that it involves a developed and naturalized level of aural thinking and, secondly, that the interaction with sound within the process commonly involves heightened levels of absorption. Schmidt Horning's (2004) analysis of the tacit skills developed by studio engineers in the first half of the twentieth century is pertinent here. She identifies aural thinking as a key way in which these engineers attuned their sensibilities towards emergent recording technologies and the particular needs of their role. Aural thinking, then, describes a developed aptitude in the identification and appraisal of sounds from within a dense sonic matrix and the ability to put them to use within a final recording. The ability to evaluate what sounds or frequencies to screen out or keep, how they will work within the overall 'aural architecture' of a recording and a tacit assessment of how they will be received by an imagined audience all became entrenched in how these professionals thought about process in their everyday working lives. Clearly, the same tuning of the ear and perceptive faculties towards set goals and aesthetic criteria are integral to the learnt modalities developed and internalized by electronic musician/producers. The primacy of sound within

[7]Personal communication with author February 2008.

all aspects of the composition process in electronic music means that the importance of developing aural thinking for producer-musicians becomes even more pronounced. It permeates all levels of their creative practice and is central to the distinct qualities of the creative experience. Through a concentrated and prolonged engagement with technology and the field, producers learn to make fast and seemingly instinctive assessments of sonic materials that are in turn judged against their knowledge of the domain with such instantaneousness that the process feels intuitive.

For example, the importance of aural thinking is evidenced in the way in which often a particular sound might be used as the starting point of the creative process. A set of frequencies might be identified from within a wider sampled soundscape and used to set off particular creative trajectories. For example, Matthew Herbert described how his assessment of the musical possibilities of field recordings was attuned to the processes and possibilities offered by digital technology but also how sonic affordances offer new trajectories within that framework:

> When you've an hour's recording to ... pick samples from ... and to create instruments from those samples, there's such detail required in making the instruments themselves that you've got to get on with that in order to enable you to actually express yourself musically rather than it being just a technical exercise ... when it comes to making music out of them you end up looking for very tiny little moments from which to pick out and amplify and one of the biggest ones being pitch; which is the hardest thing.[8]

Similarly, the Dutch producer Machienfabrik stressed the importance of listening for and identifying the aesthetic possibilities of particular moments within field recordings:

> I'm more into the texture of sounds than what the sound really is. If there's lots of stuff going around I really like to work with that. ... Maybe at first notice it might be really minimal but it's just a question of listening [back to the field recordings] and there will be lots of layers that make something interesting.[9]

Aural thinking clearly involves the intense development of listening skills where the ear is tuned towards the creative possibilities of a given sound source. To the attuned ear, open to creative possibility, distinct moments of sound come alive in terms of their creative potential. The materiality of a particular sound sample can set off differing creative trajectories and

---

[8]Personal communication with author September 2008.
[9]Personal communication with author November 2005.

also provide a set of constraints through which creative action is followed through. This particular tuning of the selective ear is, again, located at the intersection between field and affordance. The possibilities of a given sound always have to be refracted through what the producer already knows and how they perceive it will work within a given track or project.

Absorption and autotelic experience

When musicians describe tracks 'writing themselves', it suggests a fluidity of experience where the individual elements of the process come together in heightened levels of absorption. Benson (1993) sees absorption as one of the central characteristics of aesthetic experience in which there seems to be an elision of boundary between the self and object through an immersed involvement with an external artistic object. In terms of creativity this has clear parallels with Csikszentmihalyi's (1988a, 2008) concept of flow: an individual's complete mental engagement with a given activity or situation at a particular moment. Csikszentmihalyi sees this state as engendering a feeling whereby the creative act seems like an automatic or spontaneous process. For Csikszentmihalyi, flow only occurs when the individual has accumulated skill, education or tacit experience of a creative form to a level where creative action appears natural or feels as if it is taking the individual out of her everyday consciousness. As I have already argued, for the musician/producer this involves an attuning of the ear towards a balance between generic convention, creative innovation and internalized discourses of creativity, but the unity of the creative process as *an experience* means that this often defies explanation as such. Shackleton's comments in another interview clearly articulate this sense of wholeness and separation of the creative experience from the everyday:

> I can't really explain it. When I'm locked into making music I'm not thinking about anything else and it's just ... the only way I can describe it is, is that it's a compulsion, you know? Just some compulsive behavior I have that when I am in the studio and I'm making music I can't let it go until it sounds right. I just keep experimenting, and playing and changing and swapping until 'this is how it is supposed to sound'. It just comes together. Sometimes it takes a long, long time. (Quoted in Keeling 2010)

The sense of absorption within the creative experience is perhaps another reason why individuals express such a sense of personal identification with, and attachment to, the process. It has an emotional and sensual quality which McCarthy and Wright (2006, 12–17) refer to as 'feltness', that is valuable in and of itself. Csikszentmihalyi (1988) describes this as an autotelic experience, that is, an experience in which the doing itself brings autonomous gratification independently of any outside rewards. Thus the convergence of aural thinking, practical expertise and discourses of creativity within a state of flow leads to an experience which is informed by but simultaneously felt as separate from sociocultural constructs. As Ben Frost commented during the Krakow session, 'My reasons for doing it [producing music] are the same [as doing] anything I would enjoy. It's pleasure seeking. Dabbling with music and sound. It's not a set outcome, it's the getting there that I enjoy' (Ben Frost Unsound 2010). Similarly, the Canadian artist/producer Grimes describes her creative process in terms of autotelic experience and flow, commenting that she does her 'best work when I can be completely absorbed by it. ... It's always been about losing myself and losing track of time. ... It's always a good sign to realize that six hours just disappeared. ... Not being too self-aware in the process is also really important' (Lux 2013).

Experimentation and technology

However, the autotelic experience is always framed within, and mediated by, the socially constructed contexts such as those outlined in the first half of this chapter. The field is always in some way present. It is the tension and fluctuation between the autonomous felt qualities of the process and the social and institutionalized constructs of an idealized 'finished product' that simultaneously drive the creative process. As Bourdieu suggests, the 'creative project is the place of meeting and sometimes of conflict between the *intrinsic necessity of the work of art* which demands that it be continued, improved and completed, and *social pressures* which direct the work from outside' (1971, 167). The centrality of experimentation within the creative process is key here. Experimentation is simultaneously an inherent component of autotelic play, a central element in creative flow and a way of negotiating the conventions and boundaries of the field. In this way, the idea of experimentation has a profound effect on creative practices of electronic music producers in two concurrent ways. First, there is the pragmatic act of experimentation: trying differing sonic and musical elements in differing and unusual combinations, the creation of 'new' sounds through trial and error and the manipulation of existing audio material into new shapes and frequencies. Secondly, there is the discourse of experimentalism which moderates and determines how those sounds are assessed, what is used and what is discarded, what is seen as a creative choice and what is regarded as merely generic.

For the contemporary electronic musician both the field of production and the way in which creativity is experienced are fundamentally related to the use of digital technologies. As the previous chapter suggested, technology's central place as an agent within the process engenders specific ways of doing things. Particular configurations of technology offer certain affordances which are central to the autotelic experience through habitual practice. Partially, this can be connected to the nature of digital music technologies themselves. The compatibility between MIDI-driven software, control surfaces and hardware devices means that such equipment tends to become assimilated into highly personalized technological composites structured through individual ways of working and prospective aesthetic outcomes. The availability of a wide range of control surfaces which can be linked to DAWs and plug-ins in a multitude of flexible and specialized ways opens up almost limitless combinations which producers experiment with in order to find personal composites which suit their workflow and aesthetic needs.[10] Similarly, the flexibility of DAWs themselves and their compatibility with a vast array of third-party plug-ins mean that many producers construct highly personalized templates within the virtual environment of a DAW's GUI. For example, Skrillex, one of the most commercially successful EDM artists of the 2010s, points to the fact that he was not instructed how to use Ableton but learnt how to use the program through experimentation aided by the program's 'intuitiveness' which has resulted in him using it in an 'unconventional' and 'personal' way based upon trial and error with differing configurations of technology (Music Radar 2011). This finding of a comfort zone through trial and error leads to a familiarity that facilitates and enables immersion within the creative process wherein the technology feels intuitive to the level of transparency. For example, while Paul Farrier of the techno act Shadow Dancer admits that 'Getting used to [Ableton] Live was actually pretty difficult and frustrating for a long time' he now 'struggle[s] to use anything else'. He states that the process now 'feels natural' and allows 'tracks to evolve – sometimes completely by accident' (Sonic Academy 2012).

On the other hand, if they become too embedded within an individual's working practice these configurations of technology can be seen to provide *too* rigid a framework for the individual to be creative. Here, an over-familiarity with individual technologies can be seen as actually inhibiting the creative process. For example, Robin Saville of the UK electronica group Isan described the compositional process as 'more often a matter of improvisation and experimentation' through technology, but added that 'it's very easy to get into routines of "improvisation" and I often try to break those loops by deliberately not using a certain program or piece of equipment'.[11]

The fact that a perceived creative/technological deadlock was seen to be broken *through* technology is telling. It simultaneously illustrates the flexibility of the combination of technologies that make up the contemporary digital studio and the imbrication of the social and the technological within

---

[10]See, for instance, the vast array of differing software/hardware configurations evidenced in *Sound On Sound*'s regular interviews with producers and mix engineers (e.g. Tingen 2010, 2015).
[11]Personal communication with author November 2006.

day-to-day creative practices. Technology has become so central and naturalized that it is able to provide both the problem and its solution.

Such technological composites are themselves configured and shaped through the position of the individual to the cultural field in that they evolve through an assessment of the needs of a set idea of a finished end product. For example, the internationally successful DJ/producer David Guetta describes a period of experimentation and play with technology through which he could create a template which would simultaneously meet the need to make highly accessible dancefloor-orientated productions and develop progression in his sound by introducing new sonic elements:

> The way I work is to make a template, and make five or six records out of that template, and then start again. You probably can hear it in my music. So sometimes, like this last summer, I spent three months just working on sounds and templates, I didn't even touch one note. I was only trying to find new synths, new ideas, new plug-ins and new treatments for synths and sounds. (Merlino and McCrellis-Mitchell 2015)

In this instance, there is a moulding of the technological environment that can accommodate a modulation between the requirement for novelty and progression in sound and the eventual social use of the work (the dancefloor, radio, television, etc.) that is concomitant with the requirements of the institutional frameworks in which creativity takes place. For such a well-known artist as Guetta, there is a commercial imperative to maintain an established brand (through the continuance of a core sound) while introducing a discrete set of signature sounds which signify a new period in the artist's career trajectory.

Spatio-social environments

Considerations of the field are thus crucial in understanding what goes on within the techno-social-personal intersections that constitute the creative process. This extends from the wider political economy of popular music production to the specificities of its reception. An expectation of where music will be listened to (and by whom) is key to understanding how social uses might engender specific creative decisions. A particular mode of listening and an understanding of the listening environment can have material effects upon frequency range and rhythmic regularity. The relationship between an expectation of listening environment and creativity is crucial. As I have argued elsewhere (Strachan 2010), the gradual development of 'high-end' club sound systems and their increasing importance within club culture have led to aspects of the creative process within electronic dance music being grounded in a (sound) 'system led' form of production. Electronic music genres such as techno, house, drum and bass, and dubstep are primarily designed to be played in clubs through audio systems that are tuned towards increased bass response and high-end clarity. Most producers working within these genres generally also work regularly as DJs or play live in such club environments and so have a tacit understanding of how music will be experienced through particular types of mediating technology.[12] Hence, the affective power of music and the differential bodily experiences afforded by different listening environments and sound systems are elements which electronic music producers are intensely aware of as a result of their own experiences and knowledge of those contexts.

Electronic dance music producers are also acutely aware of the shared experiences through which music is put into action within those contexts: to the centrality of 'musicking', Small's (1998) widely acknowledged term to account from the multiplicity of processes and actors that go towards a given musical experience. An awareness of the crowd and how they act and react collectively within highly specific audio environments drives certain elements of creativity for these practitioners. Kreuger (2010) uses the demarcation 'social affordances' to refer to the way in which affordance is mediated and changed when it is encountered within a wider social context with multiple listeners. He argues that 'using music to construct and regulate emotional experiences and coordinate action is often a joint venture, a social practice fundamentally shaped by the shared presence of multiple perceivers' (2010, 17). With this in mind, we can see the club environment as constituting a mutual coordination of attention whereby social affordances construct and regulate how music is experienced and made sense of. Being in the company of others engenders an intensity of experience that is fundamentally different from experiencing the same piece of music in a solitary environment.

A rather self-evident connection between dancefloor, use value and creativity is a defining feature of self-declared dance musics. As a result, much of the work on dance music has represented a reflexive relationship between the creative individual (primarily the DJ) and perceivers (Langlois 1992; Firkenscher 2000). Rietveld places the relationship between the DJ and the crowd within house music as central to the practices of the genre and that 'the production of [meaning and experience within] house music is about what is at that moment the most effective on the dance floor' (Rietveld 1998, 22). The interrelatedness and mutuality of electronic dance musics with their social uses has been manifest in the structural conventions of electronic dance music ever since disco and early house music DJs used reel-to-reel tapes to re-edit and recode existing material according to the

[12]I am not suggesting that these genres are singular in engaging in system-led production at the creative/production stage. The history of sound recording has been punctuated with recordings specially designed for specific systems. Morton (2000, 40), for instance, points to specialist binaural recordings in the 1950s which were designed only to be heard through headphones.

specific needs of the dancefloor.[13] For example, Hawkins (2003, 87) notes that although the beat works as 'the basic unit of temporal measurement', the groove 'functions as a unifying unit, transporting with it a sense of regularity crossed with syncopation'. He identifies the progression of dance music tracks as working though cellular groove patterns which 'regulate different scale proportions through constantly changing patterns of repetition', providing a sense of collective movement and linearity through time and space. For Hawkins, the centrality of beat and groove regulation serve to 'encode the dynamics of club culture where the blend of identities create the impulse for expressing a wealth of shared sentiments' and thus 'feeling the beat is ... linked to a sensibility towards cultural context' and a 'simultaneous mapping of one's erotic identity onto the beat' (ibid.).

Hawkins' musicological analysis is useful as it points to how the reflexivity of the DJ–audience relationship underpins the production of dance music at a fundamental level. Further, both the relationship between frequency and the body and an understanding of the social affordances engendered by intensely shared experiences feedback into the creative process. In other words, social affordances have a key effect upon how sonic material is handled within the production process. Particular elements, including sounds, rhythms and tempos, are apprehended and evaluated according to the centrality of very particular temporary spatio-social environments within electronic dance music culture. This provides a key area where the sonic is mapped back on to the social. Perhaps the most obvious example is the use of bass within electronic dance music. As Fikenscher argues, the connection between music and the resonating body as a 'social instrument' is crucial in understanding dance music culture. He argues that 'the sonic and the kinetic spheres are incomplete parts of a form of musicking in which sound can be experienced physically, and the dancing human body acts as a musical instrument' (2000, 67). Bass offers a very defined physical affordance as it imposes itself on the body in inescapable ways and forms a key element of the erotic pleasures elicited through the interaction of the sonic and the biological. Producers within these genres are, in effect, engaged in affective work based upon a tacit understanding of how sound interacts with the body. The Dutch dubstep producer Martyn, for example, describes a highly reflexive process which is acutely informed by a direct connection between prolonged moderation of frequency and the real reactions of club crowds:

> Low frequencies do something to your muscles when you listen to them. There have been studies where low frequencies seem to contract your

[13] Reynolds (1998,16) points out 'that in the absence of fresh disco product Chicago [house] DJs had to rework the existing material into new shapes' and that 'house ... was born not as a distinct genre but as an approach to making "dead" music come alive, by cut and mix, segue, montage and other DJ tricks'.

> muscles, and because they contract your muscles you have the urge to move, you know, to get rid of that itchy feeling in your muscles. ... But basically bass is just a very physical thing. There is nothing greater than just hearing super loud bass on a good system. ... It's just all a matter of production, basically. If you know what effect a big bassline can have, you just try and make it sound as perfect as you can. I do a lot of tests. I make a track, do a mixdown of it, play it out [DJing] a couple times at different places, see what the effect is, then go back to the studio and fix it up a little bit. (Quoted in Rinehart 2008)

The way in which many electronic musicians produce differing types of music for different locations demonstrates the importance of spatio-social environment in defining the parameters of a given piece of creative work. Given the historical duality of electronic music culture since the 1980s, it is extremely common for electronic musicians to work in differing environments using differing names to indicate a firm differentiation between separate aspects of their work. Hofer (2006) uses the term 'performed multiple identities' to illustrate how electronic music producers demonstrate a clear negotiation of distinct modes of listening and performative strategies through adopting multiple personae. Producers who traverse dancefloor-orientated productions and more abstract forms intended for home listening also traverse these formal constrictions of social use and deeply embedded discourses of creativity in which progressiveness and experimentalism are highly valued. On the one hand, they demonstrate a high level of engagement with (and understanding of) a culture which has routinely sought to solicit a bodily affective response and, on the other, an embedded avant-gardism.

This can be related to a functional relationship between the management of frequency/timbre and an understanding of playback technology, social use and environment. An example can be taken of Kamal Joory, who is a Nottingham-based producer, label owner and promoter, who has worked in a variety of electronic music styles. Under the name of Geiom he has released techno and dubstep-influenced recordings and has toured various club nights and festivals related to these genres. Joory has also released work under the name of Hem to denote his more experimental electronica-based productions. In interview, he drew distinct boundaries around these differing aspects of his work with specific reference to how expected listening contexts and mediating technology had a fundamental effect upon how he approached the creative process. When asked about the crossovers between the differing aspects of his work and the incorporation of 'experimental' techniques into more dancefloor-orientated work, he responded with a discussion of frequency range and how this translated into the listening environment:

> It [dubstep] is designed to be dance music. If you make dance music there's certain factors that limit you in that you have to make music which is

> fairly regular so that it can be mixed and you have to make music which is fairly solid so it sounds good in a club. And electronica traditionally has dismissed with both of those conventions [*sic*]. So they kind of don't work together you've got to look at it in a different way. For someone like me for instance, you've got to forget about making things too tweaky … or making it sound too brittle, they don't sound good when you play them really loud. They sound wonderful if you play them from laptop speakers and you can be really intricate and granulize everything but it doesn't really work on a massive sound system. So there's things they can learn from each other but they've got to respect each other as well.[14]

This response belies how the day-to-day decisions about sound and the specificity of sonic material are mediated through a precise (and socially constructed) prism that takes into account a number of convergent factors. It suggests a stage at which aural thinking (through an understanding of, and control over, frequency range, the rhythmic/structural conventions of a particular genre, etc.) is positioned with a tacit understanding of where a sound will be heard, who will listen to it and how it will be put to use.

Summary

This convergence of factors is crucial in understanding the creative process. As analysed in this chapter, the intersection of creative practice with the social and technological contexts of popular music takes several forms. Although social use, institutional situation and spatio-social-environments provide clear constraints upon a given project, powerful socially constructed discourses of creativity have a clear effect upon how those constraints are negotiated. The position of an individual electronic music producer within a specific sociocultural framework is both something that they are acutely aware of and, crucially, often functions as a central driver in the creative process. The self-definition of an artist as *an artist* and an implicit understanding of what it means to be creative is in itself a driver for creativity. Artist-producers have an accumulated knowledge about the field and tend to internalize any social agreement as to the nature of creativity within that given field. In many popular music genre cultures, the ideas of progression and experimentation are embedded within common discursive formations that surround production. The internalization of such discourses leads to a state of creative tension between difference and similarity through which artists regularly monitor the specific choices that they make during the creative process. The compulsion to transcend the purely generic, to

[14]Personal communication with author November 2005.

produce work which is more than the sum of institutional and audience expectation, has fundamental material effects upon what producers, and indeed many other musicians, actually do.

The implications of these reflections are in keeping with broader perspectives on creativity in other fields such as science and business. Simonton (1999) places variation at the heart of an evolutionary theory of scientific creativity. In this model variation is not random, but rather subject to a process of recombination of elements of knowledge and an assessment as to their relevance to a given problem. The probability of novelty is thus mediated by the number of cognitive elements that are available to the creative individual in relation to the breadth of those elements that are deemed relevant to the problem in hand. In Negus and Pickering's (2004) terms this is what marks out certain creative artists and artistic works with a sense of 'exceptionality'. Particular individuals are skilled at manipulating the musical codes and structures that constitute the 'craft' of a particular compositional or performative tradition and possess a developed imaginative capacity. Thus, for Negus and Pickering (2004, 154), 'The painstakingly acquired creative know-how of the painter, musician or writer is drawn on to harness the stream of mental play and fantasy to certain devices and conventions while retaining the spirit of inventiveness.' As the discussion in this book up to this point has shown, digital technologies have had a profound effect upon how these processes of mental play and inventiveness are articulated, and have significantly shifted how sound is organized in many genres of popular music. Furthermore, the shifts in creative practice within the concentrated period of convergent digitization from the 1990s has had wider effects upon production values, common aesthetic conventions and generic change. The final chapter of the book therefore examines how changes in creative practice have led to differing types of digital aesthetics across emergent generic groupings and mainstream pop production.

CHAPTER FIVE

Digital aesthetics: Cyber genres, Auto-Tune and digital perfectionism

Syndrome, Liverpool, August 2014

At the opening of a set by the American electronica artist Holly Herndon, the screen at the back of the stage is dominated by the ubiquitous and familiar sight of a Facebook page. The image is the event page for that evening's club night Syndrome, one of a series of artistic events which form part of a year-long project on post-human subjectivities that I have been involved in programming and developing. As Herndon starts her set the cursor on the screen hovers over the profiles of guests who had signed up to attend the event through Facebook then clicks through to a series of individual profiles. Herndon begins to manipulate her voice through the digital audio workstation on her laptop as the audience demonstrates a mixture of amusement and discomfort at the sight of virtual public profiles being exposed in a real-world public environment. As the performance progresses Herndon's sampled snippets of her own voice form the basis of what follows. They are filtered, pitch-shifted, looped and delayed until they become seamlessly blended into her post-techno soundscapes as the personal Facebook profiles of individual audience members merge into a torrent of computer-generated imagery drawing on the detritus of hyper-mediated consumer culture.

Liverpool, January 2015

I am watching a US drama series that chronicles the ups and downs of singers, songwriters and professionals in the world of the music industry. Two of the characters are performing a duet in the living room of a domestic house accompanied by a sole acoustic guitar. While I am enjoying the craft of the songwriting and the performances, my ear is suddenly taken by something. There is a strangeness in tone as the characters begin to harmonize, a slight hint of unnaturalness and a flattening of the natural overtones of the voice. I realize that the vocal performances of the characters have been pitch corrected using the Pro Tools plug-in Auto-Tune.

These two technological manipulations of the voice are, on first evaluation, almost diametrically opposed. One is a highly self-conscious engagement with issues of identity and subjectivity relating to the technologically saturated nature of our world. The other utilizes technology in the service of realism and authenticity. One uses digital technologies to abstract the voice in an obvious manner placing it into a seemingly unnatural soundworld. The other is a 'secret' production trick, designed to be imperceptible, and employed in order to clean up otherwise engaging naturalistic performances. In short, they are utilizations of related technologies that are sonically, aesthetically and discursively divergent. Herndon's cybernetic playfulness around identity is a particularly post-digital articulation of a thematic continuum in electronic music from its beginnings in modernism and sci-fi soundtracks through to Kraftwerk, electropop, techno and glitch. Across this musical trajectory our evolving relationship with technology has been celebrated, deconstructed and placed at the heart of the recorded soundscape. In contrast, the 'hidden' use of Auto-Tune is a hangover from a realist urge within the recording of musical performance that has remained pervasive across a number of genres even as recording technologies have developed the possibility of increasingly 'virtual sonic environments' (Toynbee 2000, 70). However, the use of the technique in this instance raises an important issue relating to our ongoing relationship to technology and the enculturation of differing types of technologically mediated sound in our daily lives. The fact that the TV show's producers were confident that the level of use of the effect would be acceptable (and unnoticeable) to the bulk of the show's audience is perhaps indicative of how common the process has become. We as listeners are so accustomed to the use of Auto-Tune that its technologized sonic attributes become transparent and to a certain extent become *heard and understood as human*.

These examples illustrate that the intersection of the voice with digital technology is multiple, complex and open to differing articulations. On a broader level, what I want to suggest here is that despite their differences, these small vignettes could both be read as indicative of a post-human condition and its articulation within contemporary sonic culture. Our relationship with digital technologies has become, on the one hand, naturalized and seamless, and on the other, a mainspring of thematic and aesthetic activity. Posthumanism provides a useful point of entry to come to terms with this dual effect of digitization and digitalization as it has been a strain within critical thought constituting a 'loosely related set of ... attempts to reconceptualize the relationship between the rapidly transforming field of technology and the conditions of human embodiment' (Rutsky 1999, 24). It is not my concern in this chapter to engage with the moral and ethical debates that have characterized some aspects of debate on the subject (Hughes 2010; Bostrom and Roache 2007; Braidotti 2013) or to speculate upon the ultimate implications of computer–human relationships upon society or consciousness.[1] Rather, I am interested in how the use of digital technologies across various forms of popular music are reflective of posthumanism as a 'historical moment in which the decentring of the human by its imbrication in technical, medical, informatic, and economic networks is increasingly impossible to ignore' (Wolfe 2009, xv). Thus, a key concern of this chapter is with how the aesthetics of popular music are embedded within post-human subjectivities. How do the sounds of contemporary popular music reverberate with a context in which our experiences of the world around us and our social interactions with others are continually mediated by digital technology?

The chapter first identifies trends across postmillennial microcultures, concentrating on a number of 'cyber genres' that are simultaneously resultant from, and reflective of, the contexts of digitization. It examines how the widespread availability of computer-based production technologies has been integral to these post-digital musics since the late 1990s and how digital technologies have been crucial to their emergence and mediation. The discussion highlights how the aesthetics of these cyber genres are resonant with the post-human condition through an often highly self-conscious thematic engagement with cyberculture and the effects of digitization. The chapter then goes on to examine the digital perfectionism and the post-human voice

[1]See Nayar's (2013, 16–18) discussion of transhumanism for an overview of these debates.

in mainstream popular music. As Chapter 1 outlined, the integration of DAWs into mainstream production has led to a shift in how much popular music is made. This in turn has had an effect upon common sonic characteristics of the most heard and consumed popular music in the global marketplace. Here technologies such as Auto-Tune and particular types of plug-in synthesis and editing technologies have led to a paradigm of digital perfectionism within mainstream pop production.

This chapter, therefore, offers no claims in defining one overarching 'digital aesthetic'. Rather, it seeks to examine the textual and aesthetic implications of digital technologies across a number of generic groupings. Across the examples used in this chapter I will argue that a variety of genres and contemporary production styles are characterized by a series of signature sounds which are inherently related to the DAWs and plug-ins that are central to the creative process. Signature sounds here are indicative of the enculturation and codification of very specific sonic markers resultant from particular uses of technology. These signature sounds then become understood through the repeated use as either core sounds of a particular genre or sonic attributes imbued with particular sociocultural meanings.[2]

Cyber genres

The first half of the chapter concentrates on a number of mainly post-millennial genres (glitch, dubstep, vaporwave, hauntology and djent) whose aesthetics, growth and very existence are intrinsically bound up in the advent of digitization and the ubiquity of web culture. These 'cyber genres' have formal and expressive conventions that are simultaneously bound up with the technologies which are inherent to their creation and are reflective of the channels through which they are mediated, shared and facilitated. Often, they articulate a self-conscious discourse that places them in relation to cyberculture more generally or have a sociopolitical agenda that engages with questions of political economy concomitant with web culture. For example, as Chapter 1 illustrated, mash-up and bootleg culture, which was predominant in the first half of the 2000s, was directly engaged with issues of intellectual property brought into focus by the underlying logic of the web as a site of the free dissemination and dispersion of cultural expression and cultural texts. Similarly, the discussion that follows examines how cyber genres have generally drawn upon existing generic

[2]Brøvig-Hanssen and Danielson (2016, 2) refer to these as digital signatures, that is, musical aspects of recordings which 'bare the distinctive character of digital mediation'. Their series of in-depth close readings of recordings are revealing in terms of the musicological implications of the widespread adoption of digital techniques from the 1980s to the contemporary pop soundscape.

foundations in terms of musical convention and discourse but have been inflected by their situation in a wider cyberculture, creating new creative and aesthetic trajectories.

Self-conscious digital aestheticism

We can trace the emergence of the cyber genre as a distinct category to the late 1990s and early 2000s. This period saw the emergence of subgenres of electronic music that were highly overt and self-conscious in their use of digital technologies to forge an aesthetic which was musically and discursively rooted in an attempt to reflect the conditions and subjectivities consistent with living in the post-digital era. European and US record labels (principally Mille Plateaux, LINE, 12K and Raster Noton) began to provide a platform for artists who were developing an overtly digital minimal electronic sound, resulting in a subgenre of electronic music which became known as glitch (or microsound). Glitch as a generic marker at this stage referred to a concurrent (geographically dispersed) set of artists, including Alva Noto, Oval, Frank Bretschneider, Pan Sonic and SND, who began exploring the contrasts and connections between high and low ends of the frequency range, unwanted or glitch sounds generated from digital technology, repetition and the use of granular synthesis to develop a distinct soundworld that would have a profound effect upon electronic music over the coming decade.

Glitch and microsound were in themselves stylistic developments from the multitude of subgenres that emerged from the explosion in electronic dance music in the second half of the 1980s. Numerous producers from the early 1990s began to adapt the production techniques and many of the stylistic elements of dancefloor-orientated music in the production of more contemplative or introspective tracks. However, glitch and microsound are significant in that they constituted a distillation of the techniques and soundworlds used in previous post-techno electronica into an explicit and coherent digital aesthetic. In both sound and surrounding discourse glitch constantly referred back to the technology used in its creation and dissemination while foregrounding its position as cutting-edge digital expressivity.

Nowhere is this more apparent than in glitch's signature sounds and techniques: the manipulation/interactions of sine waves, the types of sound which are a by-product of the sound-editing process, ambient sounds associated with computer technology such as drones and electrical hums, and random and error sounds which are by-products of digital technologies more generally such as glitches, pops, hisses and CD skipping. These sounds were moulded into the rhythmic structures and timbral palette of a given track. As Sherburne (2002, 171) notes these elements 'became the *prima materia* of the music, instead of traditional synthesizer tones or samples'. As such, these elements work as technologically reflexive sonic signifiers which draw attention to their means of production and to the saturation of our daily lives by digital technology. This self-consciousness is further reinforced by the overall production aesthetic of glitch in terms of its utilization of frequency range, spatial staging, panning and use of effects. For example, Figure 5.1 is a spectrogram of a ten-second clip of Ryoji Ikeda's 'Dataplex:Microhelix' from his 2005 *Dataplex* album, which is indicative of some of the conventions of glitch's digital production aesthetic.

The spectrogram is illustrative of how glitch typically utilizes the extremes of the frequency range and has distinct forms of frequency separation. The higher end of the frequency range shows a distinct rhythmic pattern around and well above the 10 K Hz mark, frequencies which are very much in the upper range of audibility (the standard range of audible frequencies is 20–20,000 Hz). The second part of this section shows an unusually high amplitude of frequencies which go below the audible range and so can be felt rather than heard. It is a bass tone that is designed to have a bodily affect. A notable aspect of the diagram is the amount of space between high and low. Isolating these sounds in such a way draws the listener's attention to the very unusualness of their isolation and provides a physical engagement that draws into focus the relationship between technology and the body. This engagement is further accentuated by the track's use of panning. Figure 5.2 is a stereo waveform of the clip, illustrating how differing audio signals are hard panned to left and right in quick succession across the space of a bar giving the impression of rapid movement across the stereo field. The combination of frequency separation and panning is a form of extreme hyperlocalization of sound within the text. Augoyard and Torgue (2008) define hyperlocalization as a perceptive effect which is 'linked to the sporadic character of a sound source that irresistibly focalizes the listener's attention on the location of emission. When the source moves, the listener continues to follow it'. In other words, the movement of sound through the aural vista and frequency range affords an acute physical affect and a heightened engagement with spatiality.

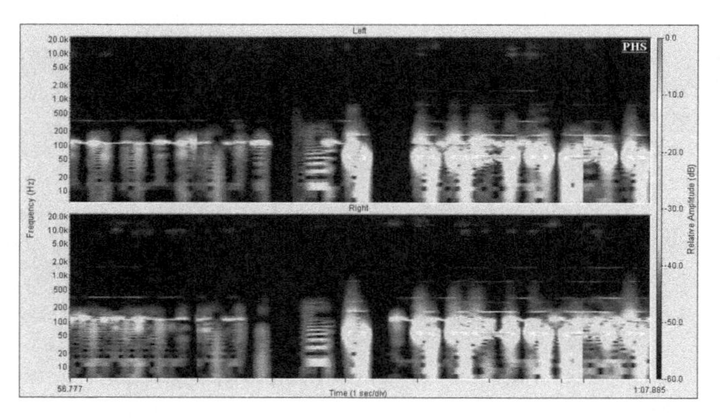

FIGURE 5.1 *Spectrogram of 'Dataplex:Microhelix' Ryoji Ikeda (2005).*

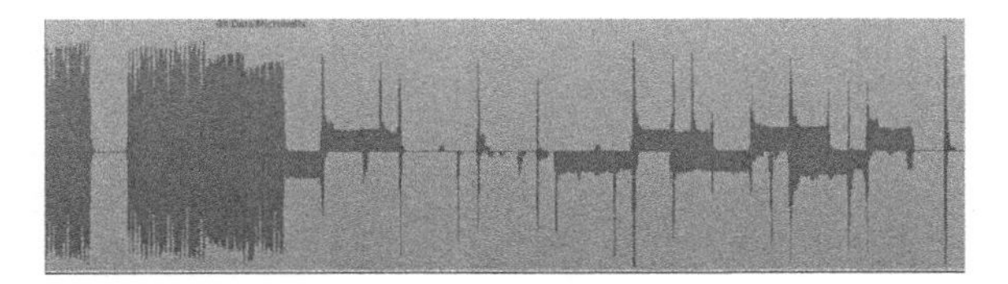

FIGURE 5.2 *Stereo waveform of 'Dataplex:Microhelix' Ryoji Ikeda (2005).*

The contrast between frequencies here also marks out the sounds' existence solely within a controlled virtual environment. The space between high and low in the spectrogram in Figure 5.1 is resultant from the recording's spatial staging, principally its lack of reverb. The middle section of the diagram is blank because there are no harmonic overtones resultant from reverberation. There is no attempt to replicate any kind of external or 'real' space and thus no attempt to 'place' the sounds in a naturalized spatiality; it is a wholly virtual staging. The track thus draws attention to its unnaturalness in a deliberate rejection of the realist discourses common across many popular music recordings. As Doyle (2005, 32) notes the use of reverb is a fundamentally 'pictorial tradition' whereby the listener is placed in relation to an imagined space, albeit analogous to space existent in the real world. It has a fundamental effect upon our perception of recordings in terms of volume, timbre and sound colouration, and is central in determining our perceptions of directionality and proximity (ibid.). Reverb has become such a normalized aspect of recordings that in many of its normative uses it is perhaps barely noticeable to the listener. The spatial conventions of glitch deliberately subvert these well-worn conventions by the use of entirely 'dry' soundworlds, the foregrounding of unnaturally gated reverbs or the placement of sonic material in clearly artificial virtual spaces.

The importance of spatiality and virtual staging within glitch's production aesthetic can clearly be situated within the post-human context. The isolation of discrete sonic elements through hyperlocalization, for instance, constitutes a conscious connection between highly stylized sonic material and the body in a theoretical environment in which debates around posthumanism have drawn attention to 'the ways in which the machine and the organic body and the human and other life forms are now more or less seamlessly articulated, mutually dependent and co-evolving' (Nayar 2013, 19). These connections are far from incidental. From its emergence, the work of glitch artists (many of whom were from art theory or philosophy backgrounds) was explicitly informed by theoretical developments that sought to examine our relationship to technology and space. Accordingly, glitch simultaneously garnered an attendant (and closely connected) poststructuralist theoretical framing: from the liner notes of various compilation CDs to its coverage in magazines such as the UK publication *The Wire* to the pages of academic journals. In these approaches the enculturation of digital technologies into our everyday lives is understood as symptomatic of a wider interpretation of deterritorialization (Deleuze and Guatarri 1983), a loosening of the relationship between culture and place as we spend more of our time engaging and interacting with virtual space. In this reading, the sounds of glitch are taken as reflecting our relationships with contemporary landscapes, both real world and virtual. For instance, Ashline (2002, 90) points out that 'microsound suggests both the inscrutability of the pointillist "click", as well as the "microtonal" possibilities of the laptop as a musical instrument, the creation of sounds within, between, and outside normative scales – sound itself as a mode of deterritorialization'. Similarly, Young (1999/2000, 47) sees glitch as 'an urban environmental music ... characterized by colossal shifts in dynamics, tone and frequency' resulting in an aesthetics which is firmly rooted within 'cybernetics of everyday life – [reflecting] the depletion of "natural" rhythms in the city experience, and in the striate plateaux of the virtual domain'.

The implications of this aesthetic have been routinely theorized as constituting an aesthetics of failure (Cascone 2004; Kelly 2009) whereby the central positioning of the sounds of error and overload invites reflection upon our relationship to technology. This has been read as either dystopian (Monroe 2003) or redemptive and progressive (Sangild 2004; Bates 2004), an aesthetic in which 'failure is turned to positivistic ends or made to "succeed" through integration into an aesthetic construct as primary content' (Hainge 2007, 35). In practice, glitch techniques were able to operate as both utopian and dystopian signifiers. This can be heard across the aestheticization of these distinctly digital sounds within the broader context of electronic music in the early 2000s. At the same time that glitch became used as a generic term it also became used more broadly to refer to a variety of practices linked to digital technology utilized in a variety of genres (e.g. the post-techno dance musics microhouse and minimal, the melodic electronica of artists such as Isan, B Fleishman, To Roccoco Rot or the ambient techno of artists such as Biosphere). Brøvig-Hanssen (2013, 172) points out that glitch techniques could be heard in a diversity of musics from hip-hop to Madonna during this period. In these contexts, glitch sounds were employed towards a variety of affective and expressive ends; funky, beautiful, relaxing, transcendent, hypnotic and so on. This range of signification is indicative of the way in which glitch was part of a wider incorporation of a natively digital sonic palette linked the personal computer into electronic music of the time.

Partially, we can relate this multiplicity of use to the particular sonic textures resultant from digital technologies and the way in which they have led to a shift in our understanding of machine noise within the contemporary day-to-day soundscape. Marstal (2002, 33) makes the point that the aesthetics of electronic music during this period saw the primacy of 'digital noise sounds' which were reflective of the fact that 'a qualitative change of everyday noise has taken place, transforming energetic and aggressive noise (factory noise, typewriter clattering) to softer and almost pleasant noise (computer humming, light sounds of copiers)'. On a broader level the mutability of digital sound is enmeshed within post-human relationships to technology.

The late twentieth century saw a shift over from a heavily industrialized economic base to a knowledge-based economy and with it a change in the sounds of everyday technologies that are predominant in our working lives. Added to this is the blurring of the lines between the technologies we use for work and leisure, social connectivity and creativity. In this context, digital technology has a variety of social uses and practices and the codification of its sonic properties within culture is therefore naturally multiple and complex.

The changes in the aestheticization of machine-generated noise during this period can be understood as a corollary to the ubiquity and multifunctionality of the personal computer in our normative activities and consciousness. Where glitch was noteworthy was in its explicitness in reflecting this relationship. Its articulation of the post-human historical moment was characterized by a constant dialogue between the theoretical and the sonic. The foregrounding of digital sound through its direct material appropriation and the placement of that sound within overtly virtual spaces within its production aesthetic unequivocally served to explore and question human–computer relationships. Glitch is also highly significant in that its overt philosophical and aesthetic underpinning was made possible by, and was directly reflective of, the DAW. Its emergence in the late 1990s directly coincided with the point in which advances in DAW design and processing speeds had meant that 'in the box' production could be done with relative ease and speed. As such, the genre signifies the emergence of a naturalized relationship between production technology and digital aesthetics that would go on to be a defining characteristic of developments in electronic music over the next decade and a half.

Dubstep

In one respect the emergence of postmillennial cyber genres is based in the pragmatic realities engendered by digitization that enabled glitch to exist. The expansion of access to digital technologies through the lowering of barriers of entry (in financial terms), an embedded design discourse of intuitiveness and ease of use explored in Chapters 1 and 2 have led to producers engaging in increasingly exploratory modalities of composition and production. The engagement with, and use of, DAWs by relatively (or in many cases completely) untrained users have meant that attempts to replicate existing production styles through the utilization of differing technologies have led to new stylistic combination and the emergence of new genres.

The emergence of dubstep in the early-to-mid-2000s can be seen in these terms. Initially, producers who would become associated with the genre were engaging with a set of stylistic continuities that were mediated, and to an extent changed, through very specific technologies. These young producers were initially attempting to replicate contemporary urban club styles (such as UK garage, drum and bass and two-step) using the technologies available to them: usually cracked software on standard home computers. Skream and Benga, two of the producers who would make some of the first releases that would first be labelled as dubstep, were at the time suburban teenage bedroom producers who were engaging with the immediate urban musical milieu through their exposure to UK garage and two-step records at the Croydon record shop Big Apple.

> Skream: We were making dubstep before it was called dubstep. Me and Benga were about 13 or 14 when we met. We started making music. It was garage but it weren't.
>
> Benga: We were following LB and Wookie and loads of other producers from two step garage. And weren't following them correctly so we went off down our own ...
>
> Skream: Yeah we went off down our own route. And we started trying to recreate these basslines. But we didn't realise is that they had like ten, fifteen grand studios and we had like, free stuff on a PC. So making these twisted basslines and it just sort of evolved from that.[3]

The adaptation of existing stylistic elements of bass music through differentiated and democratized technologies thus led to the emergence of signature sounds which were inherently related to the specific technologies used in making dubstep. Skream and Benga's use of Fruity Loops, for example, gave a distinctive sonic signature to their work, leading to elements which became key generic markers within the genre. There are various aspects of the program's functionality that can be identified here. The block system of sequencing central to the program's interface was ideally suited to the stripped down and minimal use of beats within the genre and made the layering of kick drums in order to increase the impact of low frequencies in the mix relatively quick and intuitive. In addition, the distinctive and subsequently highly influential 'wobble' bass sound (created by assigning an LFO to a Low Pass Filter as an insert effect to create a modulating wobble within the bassline) was created using the TS404, a VST plug-in bundled with the program (Hampton 2012). The distinctiveness of the plug-in can be clearly heard on the duo's early tracks such as 'The Judgement' (2003) and 'Hydro' (2004) and became a signature sound across other releases. Despite being designed as a virtual version of hardware bass units such as the Roland TB303, the TS404 did not entirely successfully replicate the analogue sounds of the original units. Instead, a relatively harsh and digitally thick sound along with the unit's built-in distortion gave an individual character to the basslines it could produce.

Early dubstep of this kind then clearly articulates particular expressions of space and identity. The characteristics of contemporary club musics work

[3]Dubstep – A Beginner's Guide – Skream and Benga Full Interview (Part 1), https://www.youtube.com/watch?v=m8kujaBMQQU (accessed 28th February 2016).

as sonic signifiers of urbanity and youth that were mediated through the private space of the teenage bedroom and a DIY aesthetic of self-production. This is not to claim that dubstep was not rooted in the sociocultural contexts that have been central to the emergence of electronic dance music genres throughout its history. To a large extent the emergence of dubstep can be clearly related to particular articulations of physical space and place; the club cultures of London (specifically, FWD>> at Plastic People, and DMZ at Mass) and an emergent network of DJs producers, pirate radio station operators and promoters which were located and operated in this physical space. Many of the key operators in the emergence of the scene in Croydon were either school friends, had family connections or met through the Big Apple record shop (Martin 2015). In addition, clubs were important as spaces where a variety of producers from across London working in post-UK Garage styles became aware of each other's work and a competitive and collaborative scene began to emerge (ibid.). In this sense the emergence of dubstep was in keeping with how localized scenes based around central institutions, real-life social networks and communities of interest have been crucial in the stylistic development of genres across the history of popular music (Straw 1991; Kruse 2003).

However, the way in which the genre spread and evolved both within underground culture and through its incorporation in the mainstream can be firmly situated within the virtual realm. For example, dubstep DJ sets broadcast by locally based pirate radio stations such as Rinse FM (Croydon) had an instantaneous and profound effect through their sharing via the internet. Almost immediately after the genre cohered in south London, Rinse FM's broadcasts found an international audience across Europe and the United States. Initially, the shows (along with dubstep DJ mixes from other sources) were being recorded and shared through the work of a site set up in 2006 by Deapoh, a dubstep enthusiast and Big Apple regular with access to an internet server via the domain name Barefiles.com (Municiple 2006). This meant that shows could be heard internationally within hours of their initial broadcast. At the same time, sites such as Dubplate.net were streaming new dubstep tracks (albeit at low resolution) as soon as they were released. These sites, combined with a highly active web community through portals such as Getdarker, Dubstepforum and Hyperdub, led to an acceleration of producers engaging with the style on an international level at a relatively early stage in the genre's development. By 2006 the genre had been adopted by US producers and DJs to an extent where many of the major urban centres had regular club nights (McKinnon 2007). The fact that the signature sounds of dubstep were resultant from the use of easily available DAWs (and the fact that many young dubstep producers were uncharacteristically candid about their techniques) meant that the stylistic elements of the genre were easily copied, replicated and adapted across a variety of international locations by producers unconnected with the physical spaces of the scene through which the genre had emerged.

Dubstep was not the first UK urban scene to have free or cracked DAWs at the heart of its creative techniques and production values. At the same time grime (an almost 'entirely a DAW [based] phenomenon' Jones 2012, 187) had taken a DIY approach to the perceived commercialization of UK garage through teenage artists and production teams such as Skepta and Ruff Sqwad using Fruity Loops to create raw, stripped down productions that would become definitive within the genre (Hancox 2012). The software (in its later incarnation as FL Studio) would remain central to grime through its second wave in the 2010s. Again, the ease of use of the program's interface coupled with its centrality in the original grime sound led to the continuing perpetuation of the genre in the urban settings from which it had first emerged. Second-wave grime producer Darq E Freaker, for instance, comments that during his school days 'everyone had Fruity Loops on their computers at home and making tunes was more like a game' (Fintoni 2015). However, the highly localized nature of grime through its heavily accented MCing and foregrounded representation of London street culture, often linked to specific London postcodes or 'ends' (Ilan 2012), meant that it was less easily transportable than dubstep. The instrumental nature of dubstep along with its cultural appropriation from a wide variety of sources (electronic dance musics, Jamaican dub, Eastern musics) meant that it was a hybridized, cosmopolitan and easily transferable genre from its inception.

While dubstep's emergence was clearly in keeping with the stylistic cohesion of past electronic music genres, it is its rapid internationalization and the manner in which this took place that is significant. While less self-consciously reflective of post-digital contexts than glitch or subsequent cyber genres in terms of its thematic orientation, the mediation and spread of dubstep is significant in that it would become central in the way that a whole raft of subsequent electronic music genres would emerge and cohere. Communication technologies allowed for the rapid global spread of the genre and the widespread availability of its production technologies facilitated the loosening of centrality of place in its production contexts. By the time that dubstep reached its commercial zenith at the end of the decade, US producers were leading the way in its crossover to a much wider market. This period also saw the emergence of almost wholly placeless virtual scenes that were entirely situated within the virtual realm.

Hauntology, vaporwave and the endless digital archive

Dubstep's international mediation and exponential growth through web-based channels was followed from the late 2000s by a proliferation of sometimes short-lived cyber genres whose emergence was entirely facilitated through internet communities and new media gatekeepers such as blogs and online magazines. The naming of new genres, the discussion of their aesthetics, discourses and formal attributes often took place through strategic groupings by influential blog writers or discussion threads on relevant internet forums. Stylistic groupings such as witch house, hauntology, chill wave, vaporwave and PC music were all resultant from the public discursive spaces of the internet. These genres were also characterized by a thematic and aesthetic engagement with the wider cultural effects of digitization. In contrast to glitch's central articulation of a hyper-modern thematic, the creative thread running through these genres is a concern with, and mining of, textual and aesthetic elements of often esoteric moments from popular cultural history.

For example, hauntology and vaporwave draw upon the internet's almost infinite scope as a textual repository of media and music from across differing eras, constructing a hyper-real version of the past. Sexton (2012) sees the label Ghost Box (one of the most prominent labels within the hauntology movement) as being engaged in particular constructions of cultural memory enmeshed in broader forms of curation, heritage construction and collecting facilitated by the archival possibilities of the internet through sites such as YouTube. The label's construction of an at-times whimsical, sinister and often comic Englishness set in an early 1970s parallel universe is simultaneously nostalgic and post-digital. The juxtaposition of the sounds of early electronic music and educational film soundtracks placed into often strange, non-linear narratives both constructs a new audio world and highlights the strangeness of the esoterica to be found at the click of a mouse. Similarly, Reynolds (2010, 84) points to YouTube providing 'opportunities for ... time travel tourist trips into exotic pockets of cultural strangeness' for contemporary electronic artists. Vaporwave's aesthetic is built upon the reworking and recoding of textual materials culled from YouTube from the 1980s and early 1990s with a specific emphasis on advertising and demonstration material for consumer computer products.

This engagement with discourses of memory is often undertaken through a nostalgic, yet knowing and ironic, appropriation of the sonic elements of past genres mediated through contemporary production techniques. For example, both vaporwave and hauntology employ a set of similar strategies in their evocation of memory. First, there is an appropriation of signature sounds (particularly synthesizer sounds, electronic drum kits, string pads, digital piano and bell sounds) from previous eras. For example, James Ferraro's foundational vaporwave album 'Far Side Virtual' uses FM synthesis and early digital technologies to evoke various 'lost' strains in 1980s culture (New Age music, corporate videos, advertising) while staples of the Ghost Box label such as Focus Group and Belbury Poly have analogue synthesizer technology at the heart of their formation of a virtual 1970s whimsy. Secondly, there is a highly explicit use of samples, often rendered in a cut-up style, whereby the obviousness of the displacement of the original sound source is highly self-conscious and self-apparent. For example, from two of the albums which were instrumental in the emergence of vaporwave (*Chuck Person's Eccojams Vol. 1*, produced in 2010 by Daniel Lopatin, one of the international stars of experimental electronica in the 2010s, and Macintosh Plus' 2011 *Floral Shoppe*) to more recent releases (such as Nmesh's 2014 *Dream Sequins®*), the genre has been characterized by work based around short but recognizable samples drawing on an eclectic range of sources from mainstream 1980s and 1990s pop hits, soul and R&B of the 1980s, through to New-Age synth music which are looped, slowed down, filtered and drenched in reverb or digital echo. The use of samples in this way is indicative of a deconstructive or abstracting mode of the materialist process (Demers 2010, 59–63) of composition within electronic music more generally whereby a play on the associations of the original sample works towards its semiotic repositioning. In both hauntology and vaporwave through the frequent use of cut-ups, the mode of sampling draws attention to its own assemblage and ultimately the materiality of cultural memory in the digital age. They are part of what Brøvig-Hanssen (2013, 159) calls an aesthetic of 'opaque mediation' whereby through the obviousness or staging of a sample 'a listener's focus is ... directed not only toward what is mediated but also toward the act of mediation itself'.

The cut-up mode serves to reify the past into highly stylized audio segments, which through their repetition and effect staging become uncanny in their detachment from their original context. For example, there is an extensive use of delay and echo across the genres that further rework the original source materials in terms of memory. Echo and delay mediate such apparently retro soundworlds by drawing on culturally constructed associations that render them transformed. As Veal (2007, 198) points out, within 'the sonic culture of humans, the sensation of echo is closely associated with the cognitive function of memory and the evocation of the chronological past' while simultaneously evoking 'the vastness of outer space, and hence (by association) our chronological future'. At the same time heavy reverberation has well-worn cultural associations relating to the acousmatic voice, disembodied from its source evoking a sense of unease and unknowability in the perceiver (Labelle 2010). This multiple signification simultaneously brings to mind a shared cultural memory, a critique and celebration of the utopianism of past technologies and a reflection upon processes of remembrance in the context of post-digital mediation.

It is the way in which these materials are reworked that lends them a self-reflexivity that constantly draws attention to their materiality. Aside from the extensive use of delay and reverb there is a common use of modulation effects which are directly redolent of the materiality of the physical media of bygone ages of popular culture. Uses of extreme chorus and phasing replicate the sonic effects of the technological malfunction of formats such as audiotape or VHS videocassette due to ageing, overplaying or demagnetization. Similarly, sources such as 1980s synth pads and electric pianos are processed through effects such as side chain compression, a process which gives a rhythmically pumping and woozy feel (see, for example, Tycho's 2013 track 'Ascension' or 2814's '恢复' from 2015). Side chain compression serves to puncture the sonic skin of the original sound, giving it new rhythmic emphasis by using a signature sound of recent electronic dance music. The utilization of these types of effect essentially works as distancing devices that further serve to highlight the deconstruction of the original source materials. Any sense of nostalgia evoked by the musical materials and sounds of the past is mediated and ironized by an unmistakable sonic patina that is central in their recontextualization.

In a revealing interview with Reynolds (2010, 83) Daniel Lopatin positions his work as not being driven by nostalgia but rather by a critique of the discourse of linear progress that drives capitalism and consumption. He bemoans the fact that the 'rapid-fire pace of capitalism is destroying our relationships to objects' and that his mining of popular cultural memory is therefore an articulation of a 'desire to connect, not to relive ... we homage the past to mourn, to celebrate'. Such a critique of capitalism is within the continuum of a series of strategic articulations within alternative music and culture over a number of years. For example, Vale and Juno (1993, 4) make explicit links between the appropriation of 'forgotten' mass culture, such as exotica and lounge music, and an opposition to *contemporary* mass culture. For Vale and Juno it is its very obsolescence in terms of the mass market that marks its appeal to a new audience.

> Society uses aesthetics in order to control us through our buying patterns, and to coerce us into buying higher priced commodities. ... Therefore, it's a subversive act to rediscover and value that which is cheap and readily available ... what society has thrown out. (ibid.)

Such reasoning is typical of how past musics are commonly historicized with regard to contemporary sensibilities. There is something distinct about the explicit engagement with technology that makes genres such as vaporwave and hauntology resolutely post-digital articulations of this continuum. First, the overabundance of actual historical media and content in the internet's endless digital archive elicits a new relationship with the past. Secondly, the capturing of the past through digital media means it is easily manipulated through digital editing technologies and invites engagement with its own materiality in terms of sound and image quality. Both have an engagement with what Reynolds (2010, 81) calls a 'buried utopianism within capitalist commodities, especially those related to consumer technology in the computing and audio/video entertainment area'. The contemporary gaze upon the material qualities of aural articulations of modernity from the past renders a clear sense of irony to this utopianism. The appropriation of sound and sonic textures from previous production eras gives a precise semiotic association with the time and sensibility of a given historical moment. In this case, sounds which were redolent with a sense of utopianism or past articulations of modernity are ironized and deconstructed through their re-contextualization with a post-digital soundscape. They are an example of what Ezra (2014) calls the 'reuse economy', a resistant re-appropriation of cultural objects that imbues a double-edged value. In this theorization 'reused objects intermittently flash "then/now," "then/now," signifying two eras at once: that in which they were manufactured ... and the era in which they are being reused' (2014, 380) in a rejection of hypercapitalism's logic of built-in obsolescence. For Ezra, reuse serves a dual function which 'privileges resistance and underlines the posthuman dissolution of the boundaries between people and objects' (Ezra 2014, 379).

Reuse and contemporary constructions of cultural memory are, I think, important in understanding the utilization of digital technologies in the aesthetics of contemporary electronic music. The examples outlined here are characteristic of a broader concern with memory and materiality. A range of highly influential artists (from the post dubstep of Burial and much of the output of the Triangle Records roster to the contemporary electronica of Oneotrixpointnever, Dean Blunt and Tim Hecker) over the past decade have used a similar set of techniques (the spatiality and reverberation of dub, the physical degrading of sound through distortion, granular synthesis and filtering) to invoke a similar set of discursive concerns and utilize a similar set of affective strategies. This broader aesthetic is significant in that it is reflective of the material contexts through which memory is stored and articulated. While subjectivity and memory have always been to an extent post-human, in that they are dispersed across the embodied and the wider material or technological domain (Hayles 1999), the embeddedness of our subjectivities in the virtual has engendered new, accelerated formations of this relationship. The archival is now a fundamental organizing strategy of HCIs through the 'archival metaphors' embedded within GUIs, the archival trail of social media, the construction of the individual as a 'miniarchivist' through cloud and mobile storage (Parikka 2013, 2) and the boundless dispersal and availability engendered by the web as a vessel for collective memory. The reuse and recontextualization of the remnants of past sonic culture through digital technologies thereby simultaneously highlight their status as materials of memory and the materiality of the remembrance process in a virtually saturated world.

Djent

While there is a high level of self-reflexivity in what I have demarcated as cyber genres so far, the characteristics are applicable to a range of microcultures that have emerged in the past decade and a half. The integration of DAW-specific techniques within emerging creative trajectories and sonic conventions is also discernable in subgenres related to long-established styles that are not self-apparently electronic in nature. For example, around 2010 djent began to emerge as a subgenre of metal. Taking musical cues (such as palm muted, high-gain drenched riffs played on downtuned guitars, shifts in time signatures and high levels of instrumental dexterity) from previous technical metal acts Meshuggah and Sikth, solo musician producers, and bands such as Periphery, Animals as Leaders and Paul Oritz began to hone a sound that was simultaneously redolent of existing subgenres of metal and resolutely digital in overall feel. Djent is significant in that it is a clear digital offshoot of a culture that has been based around liveness, virtuosity and authenticity but selectively adapts and evolves aesthetic elements of that culture in response to new technologies. The genre's social and cultural trajectory from within cyberculture and the centrality of digital technologies within its creative processes have had a fundamental effect upon its aesthetic qualities. First, djent as a genre is inherently rooted within online culture. Many of the acts that became associated with the sound were actively involved in message boards centred around technical metal music and home recording techniques. The sharing of information and demo recordings within these forums created a network of bedroom metal producers and an initial buzz in the online community which enabled a number of bands to develop an audience and go on to have sustainable careers. As Sander Dieleman, webmaster of got-djent.com (web portal and online hub for the djent community), explained in an interview with the *Guardian*, djent can be understood as part of the proliferation of online musical cultures that emerged in the wake of digitization regarding the genre as an almost wholly 'online phenomenon' facilitated by the technology's capacity 'to produce professional-sounding music in your bedroom. If you want to play djent, all you need is a guitar, a computer, a guitar interface and understanding neighbours' (Thompson 2011).

One of the key markers of the genre is a guitar sound primarily associated with Line 6's Pod XT amp modeler, a digital sound module which was produced in order to simulate differing types of guitar amplifier in one unit. Initially borne of necessity, as it was designed for guitars to be directly inputted into computer soundcards, in the context of djent, the virtual or replicative nature of such a digital technology becomes secondary to its ability to create signature sounds. The unit is able to produce the thick distortion reminiscent of a high-gain amplifier that is simultaneously highly controllable and tweakable in terms of frequency range. The subsequent effect gives a clarity in the individual tracking of guitars and allows emphasis to the staccato nature of riffs within the genre and the precise technical guitar work.

DAW technologies enable these artists to pursue an often individual form of composition and production completely at odds with how metal has been traditionally worked on in a collective process of creativity. In addition, djent explicitly and unashamedly exploits the full possibilities of the DAW within its creative process and production style. For example, Misha Mansoor, guitarist and producer with Periphery, outlines a complex production process in which the band's drummer plays the parts on a Roland TD20 MIDI kit after which the MIDI information is used as the basis for the drum tracks which eventually appear on the final recordings. The actual drum sounds are composed of samples within Toontrack's program Superior Drummer:

> So we were able to get Matt's unique performance, which I'm really happy with, and then replace it with the Superior Drummer samples. I feel that's the best of both worlds – you get the performance and you get the sounds, which in my opinion are very hard to match without going to a ridiculous studio, because those are recorded in the best studios in the world. I'm very happy with the way the drums came out on the album as a result. That is probably the way we're going to continue to record. He was more than happy to play through an electronic kit, and that way we have all his beats and his fills and ghost notes. (Chopik 2010)

The resultant production aesthetic resultant from djent's combination of signature sound and DAW-based editing capabilities is one of digitally mediated precision. Recordings are characterized by a high level of separation in frequency range, clarity of individual tracks (which allows for an unrestrained foregrounding of virtuosity) and, in some cases, an incorporation of signature sounds and techniques from electronic dance music (filtering, keyboard pads and arpeggios). This specific combination of strategies is indicative of a digitally native microculture led by predominantly young musicians and producers at ease with all aspects of contemporary technology at odds with pervading discourses of rock. Rock and metal have traditionally been grounded in discursive constructions in which recording technology has been regarded in ambivalent terms at best. Recording technologies have been viewed as transparent and realist, a means to an end which supposedly is there to capture a sense of liveness. As Zargorski-Thomas (2013, 206–8) argues there has always been a negotiation between performance authenticity and recording practice within rock music. As a corollary from this position there has been a resistant strain in rock and metal discourse in which recording and production are viewed as a mediating process with the power to deceive or trick a listener (take, for example, the common practice from the 1970s and 1980s for live hard rock and metal albums to explicitly state that they contained 'no overdubs'). While djent clearly foregrounds the virtuosity which has been central to metal (Walser 1993), it is a virtuosity which is self-apparent and centrally facilitated by the genre's production conventions.

Djent's embrace of technology constitutes a discursive acceptance of the dispersed nature of creativity across the biological/technological strata. The embeddedness and unselfconsciousness of this position belies a dual consequence of the digital turn. First, it is indicative of how access to digital technology has changed the nature of what it means to be a musician across all types of popular music practice (highly engaged in production with a fluid and tacit engagement with a divergent set of technologies). Secondly, it signifies a more general sense of comfort with digital technologies, naturalized within a generation that has grown up within the sociocultural wake of digitization.

Hence, while less overtly engaged with the conditions of posthumanism than the electronic genres outlined previously, djent is nonetheless deeply grounded in the post-human historical moment and the accelerated effect upon subjectivity facilitated by our everyday engagement with the virtual.

The DAW and digital perfectionism

The highly composite mode of creativity found in djent is indicative of a new normativity in the creative process across popular music genres. The ubiquity of the DAW within recording practices across the production of popular music more generally, as evidenced in Chapter 1, has had a profound effect upon both how recordings are made and the sonic characteristics commonly found in recordings. This applies to music that is not only demarcated as explicitly electronic, but pervades almost all areas of recording across popular music genres. First, the level of editing and the ease of use in editing facilities afforded by programs such as Logic and Pro Tools have meant that recordings tend to be made up of highly composite elements. The fact that these virtual environments afford much easier and quicker ways of cutting, pasting and blending audio material has led to their almost ubiquitous use in the modern studio setting. Although these techniques have been integral to the production process since the advent of tape, the fact that they are naturalized into the GUIs and functionality of DAWs has led to their centrality in modern recording and production. The practice of compiling (or comping) parts of numerous elements of individual instrumental or vocal takes into finalized tracks is now 'standard practice in professional productions' (Senior 2011, 107) and so commonplace that it is now an essential skill that a large proportion of producers would use on a daily basis.

Secondly, the fact that DAWs functionality is also based around the use of plug-ins has meant that time stretching and pitch correction have become commonplace within the way which recordings are made. For example, plug-ins and built-in functions of DAWS, such as Pro Tools' Beat Detective or Elastic Audio, Digital Performer's Beat-Detection Engine or Logic's time-stretching feature, allow for a level of editing, rhythmic precision and dynamic consistency in recordings of acoustic drums and percussion that had hitherto been the preserve of electronic drum machines and sequenced sample-based percussion tracks. Zagorski-Thomas (2014) sees this in terms of the visual and functional affordances of DAWs creating something akin to a will to perfection. Seeing the affordances of DAWs as a form of script (Akrich and Latour 1992), he argues that visual affordances engender a process of 'atomising the act of composition and record production … exposing every aspect of performance to closer and closer scrutiny' resulting in a 'clinical quest for technical perfection' and that ultimately 'as soon as the technology to fix blemishes exists … the pressure to utilize is brought to bear' (2014, 135).

This has also been seen to have implications in terms of the types of performances that are required within the recording process and resultant authenticity paradigms. Milner (2009, 340) points out that the retriggering of elements of drum performances to maintain consistency (in which the recorded sounds of kick, snare drums, etc., are replaced with samples through the use of such plug-ins) is such a widespread practice that it negates the need for dynamically (and rhythmically) precise performances in the tracking process. While Milner's assessment is somewhat mournful of traditional musical skill, Zagorski-Thomas (2013) describes the current ubiquity in the recording process of non-linear recording, quantizing and editing techniques within drum tracks in slightly different terms. Rather than constituting a degradation of musicianship, he argues that these technologies are situated in 'an alternative form of authenticity' prevalent among musicians 'based on an understanding of the techniques and processes of record production' that 'recorded music requires a different approach to performance practice than concert performance'.

Whatever the authenticity debates surrounding the use of these plug-ins, they are indicative of how the internal functionality of DAWs afford a level of precision and ultimately a digitally constructed notion of perfection in contemporary recordings. The capabilities of the technologies themselves are integrated into an understanding of how contemporary recordings *should* sound. Zagorski-Thomas' (2013) conclusions are, I think, pertinent and significant in this respect. The acceptance that the 'performances' that ultimately end up on a recording are a combination of bodily, human action and computer-based functionality signifies a deeply enculturated relationship between the two. The implications of this relationship are even more striking when we consider another key functional relationship engendered by VST plug-ins: pitch correction and the human voice. There is something so fundamental in our understanding of the voice as a cultural marker of our humanity and individuality that its contemporary relationships with technology in the recording process are worth unpacking at length here. The remainder of the chapter, therefore, examines the history of use of the VST plug-in Auto-Tune over a number of differing genres, arguing that it is redolent of a diverse set of post-human relationships and subjectivities.

Pitch correction and vocal staging

Although pitch correction had been part of studio production for some time, the launch of Antares Audio Technologies' Auto-Tune in 1997 made the process widely accessible and easily usable as a DAW plug-in across platforms. Auto-Tune is a pitch correction plug-in initially developed for Pro Tools that allows real-time pitch correction without changing the speed at playback.[4] Envisaged as an editing tool that would subtly allow producers to correct pitching inconsistencies in vocal performances, the plug-in has had wide-reaching implications across a multitude of genres since its launch. On one level, the 'overuse' of the tool has led to a signature sound related to a pronounced vocal effect. On the other, its normative use has had pervasive yet less explicit consequences for accepted norms with regard to the sonic attributes of the voice in pop production more generally. These two consequences constitute a shift in what Lacasse (2000) calls 'vocal staging'; that is, any 'deliberate practice whose aim is to enhance a vocal sound, alter its timbre, or present it in a given spatial and/or temporal configuration with the help of any mechanical or electrical process ... in order to produce some effect on potential or actual listeners' (2000, 4).

The differing effects of Auto-Tune are related to how the tool is used and the level to which the voice is manipulated. Setting the key of a song within the plug-in allows Auto-Tune to analyse sections of a vocal line in order to move 'wrong' or imperfectly pitched notes up or down to an intended pitch within a given user-defined scale. The effect can be nuanced in terms of timing and transition between pitches, both of which have a clear effect upon how the sound is perceived. This is largely to do with quantization, or how acutely the effect is made to fit with a predetermined grid relating to pitch in terms of timing. In the 'adjustment time' of the plug-in any setting below 15 begins to have a pronounced effect and setting the parameter to zero produces an acute digitization of the voice. Heavy quantization erases naturally occurring glissandos between notes, making the transitions between them immediate. This gives the perception of a highly processed sounding timbre that has often been understood as robotic or inhuman. In normal operation, when quantizing is less pronounced, Auto-Tune allows for a more 'natural' sounding pitch correction whereby the microtonal shifts remain perceptible.

We can think of the differing levels of use of Auto-Tune as constituting 'explicit' and 'intrinsic' usages. Explicit uses clearly work towards the othering of the voice as 'unnatural' or 'processed', thereby drawing upon a historically embedded and emergent set of semiotic associations. While intrinsic usages are intended to be hidden or at least unobtrusive, they nevertheless often produce particular audible qualities in the voice. They are forms of 'vocal setting', that is, specific configurations of vocal staging which have distinct characteristics 'in terms of loudness, timbral quality, and spatial and temporal configuration' (Lacasse 2000, 5). Auto-Tune was envisaged as a transparent technology and its explicit usage is essentially

---

[4]It should be noted that there are a variety of pitch correction plug-ins that followed Auto-Tune into the marketplace such as Melodyne and Waves Tune. Given its centrality and precedence in the market and its use within popular accounts and discussion, Auto-Tune is used as a catch-all term for vocal pitch correction within this chapter.

a misuse or overuse of the function for which it was originally intended.[5] However, the widespread adoption of explicit use has led to Antares themselves positioning their product in terms of dual functionality in terms of marketing and use. For example, the introduction of the manual for the product makes a clear distinction between 'pitch correction' and the 'Auto-Tune vocal effect' or pitch quantization even though they utilize the exact same audio processing (Antares 2011, 5).

Explicit uses of Auto-Tune

In one way, we can see the explicit use of Auto-Tune within a tradition of technologically mediated and synthesized voices within popular music stretching back to the widespread use of the vocoder from the 1970s. The overtly 'robotic' processing of the voice via the vocoder in electronic music genres such as disco, funk, electro and certain types of avant-garde rock music have been read as representative of a strain of futurism within popular music (Dickinson 2001, 334). For example, in a convincing analysis of the use of the vocoder in the work of Kraftwerk, Biddle (2004, 82–3) points to the use of the technology by the group as being manifest in four main ways; 'ironic displacement' or a self-conscious 'flattening' of the emotional dynamics common in the expressive language of the voice, 'dehumanization' through the use of automaton or industrial (sonic) imagery, a 'critical disengagement' or rejection of dominant paradigms of the organic or authentic embedded within many popular music genre cultures and a 'refashioning of … models of creativity' which are reflective of the increasing automation of daily life and a subsequent cybernetic turn in subjectivity. While common sense would imply that, as a highly processed and technologically overt category of vocal staging, the explicit use of Auto-Tune is congruent with such thematic articulations of alienation, mechanization and futurism, I want to suggest here that the enculturation of the effect across the soundscape of contemporary production has resulted in a more diffuse and complex set of meanings which are reflective of the multifaceted and naturalized character of contemporary post-human subjectivities.

The explicit use of Auto-Tune as a stylistic production marker can be traced to Cher's international hit 'Believe', which was released just a year after Antares released the plug-in. Although the record's production team initially indicated that the effect had been achieved through the use of a

---

[5]The history of popular music is, of course, punctuated by the employment of specific technologies in ways which deviate from their designed purpose. Julien (1999) refers to this as a 'diverting' of technology, citing the specific example of double-tracking vocals whereby the slight phasing caused by the technique of using two vocal takes to 'thicken up' a vocal performance became valued as a signature sound in the work of the Beatles.

Digitech Talker (a stompbox style vocal synthesizer), the distinct vocal style afforded by the use of pitch quantization was quickly recognized and adopted by other pop producers and became known as the 'Cher effect'. In the immediate years following, explicit pitch quantization became a trend within mainstream pop music and could be heard across the range of then-mainstream genres, from commercial dance music (e.g. Eifel65's 'Blue (Da Ba Dee)' 1999, Daft Punk's 'One More Time' 2000) to huge selling pop acts such as *Nsync's 'Pop' 2000, True Steppers Feat Dane Bowers & Victoria Beckham's 'Out Of Your Mind' 2000. The success of these records led to the adoption of explicit usage of Auto-Tune in order to lend patina of contemporary pop sensibility to a variety of productions. As Dickinson noted as early as 2001 the technique had become 'one of the safest, maybe laziest ways of guaranteeing chart success' (2001, 333). For example, Faith Hill's 2000 hit 'The Way You Love Me' explicit use of the effect in the song's chorus was clearly a strategy which was part of Warner Brothers' campaign to market Hill as crossover artist who transcended the core country music audience. In a US radio market so clearly segmented by genre and its connection to perceived demographics, the use of explicit Auto-Tune here was clearly part of a set of production decisions designed to expose the artist to a wider audience. The effect was fitted with a then-contemporary production aesthetic that chimed with the contemporary soundscape of pop radio, clearly lending the track a marketable and modern sonic signature.

We can see these examples from the late 1990s and early 2000s as being the start of a process of enculturation, whereby the explicit use of Auto-Tune becomes normalized and ubiquitous. The longevity of the technique in mainstream pop production is revealing in terms of the dominance of a new digital aesthetic within the soundworld of pop music. Rather than being a mere short-lived novelty, the explicit use of Auto-Tune has had a shelf life that has seen it become something of a signature sound within contemporary popular music. By the end of the 2000s the technique was so ubiquitous as to be beneath comment. For example, two of the biggest international hits of 2009, Taio Cruz's 'Dynamite', the Black Eyed Peas 'I got a feeling' and the bestselling international single of 2010, Ke$ha's 'Tik Tok', all make extensive use of explicit Auto-Tune. By this time the technique had essentially become a major device in the producers armoury of tools that could be utilized in creating an overall sound of the recording. Rather than being overtly thematically grounded in any sense of the robotic or evoking alienation or the futuristic, the vocal staging of all three tracks is perhaps more used to connote an aspirational, contemporary, youthful, vibrant and euphoric feel. The flattening out of the individual characteristics of human voice results in a homogeneity of tone. As Pettiman (2012, 151) argues, Auto-Tune has the capacity to 'aurally airbrush … any vestige of singularity out of the voice of the performer so that the sonic performance becomes … a whatever-vox: a voice without qualities (or rather, a voice displaying the homogenized quality shared by all other artists who utilize the same software)'. What is perhaps left is a placeless hyperreal address which chimes with a sense of globalized aspiration, reflective of global capitalism's cross territorial marketing of brands, products and services. As Von Appen (2015, 48) notes these tracks are part of a dominant strain in contemporary pop music in which its explicit cultural values 'promote hedonism, individualism, physical attraction, luxury, big egos, getting blasted on weekends and rather conventional heteronormative gender stereotypes'. As such they are reflective of a constructed idea of what the major entertainment industries are selling and constructing as the 'the ideals and yearnings of today's … teenagers' (ibid.).

The centrality of explicit uses of Auto-Tune in pop should also be read in the concurrent and related strain in various musics of the Black Diaspora. Throughout the first decade of the 2000s explicit Auto-Tune use became pervasive across a number of such genres. R&B, Dancehall and hip-hop artists in particular began to explicitly foreground the technique in the construction of their signature sounds. Tanto Metro & Devonte's 2001 dancehall hit 'Give It To Her' ushered in not only a subsequent plethora of similarly Auto-Tuned dancehall tracks but also a defining mode of the effect's use across Black Diasporic musics. First, the track uses the 'zero setting' over the entirety of Devonte's vocal track in the chorus. This was in contrast to most previous uses in pop music (aside from overtly 'robotic' uses such Daft Punk and Eifel65) where the technique was used to create sonic hooks at particular moments (usually in the chorus). Secondly, the track uses the effect to produce a digital warble directly resultant from the plug-in not being able to process the rapid melismatic shifts characteristic of Devonte's soulful tenor vocals. Both the extended use of Auto-Tune and the digitized melisma worked their way to the centre of the aesthetics of core diasporic musics over the next decade. Perhaps the most influential factor in this trajectory was the 2005 emergence of the US hip-hop star T-Pain. After scoring a top 10 Billboard hit with 'I'm Sprung' (in which both verses and choruses were heavily Auto-Tuned) T-Pain went on to score several platinum selling albums in which Auto-Tune was used across the vocal staging. T-Pain became so associated with Auto-Tune that he released a series of cash-in products such as the 'I am T-Pain' toy microphone and app, both of which allowed users to replicate the effect (Feifer 2012). The technique was subsequently adopted by a number of high-profile hip-hop stars, including Kanye West, Lil Wayne and Snoop Dogg, and has since pervaded the soundworld of contemporary hip-hop so profoundly that it has become a key style indicator within the genre.

It is tempting to see the explicit use of Auto-Tune in African American musical cultures as being in a tradition of Afrofuturism; a strategic series of appropriations of technology by black American artists in a very self-conscious engagement with identity issues. Nelson (2007 in Rollefson 2008) defines Afrofuturism within jazz, funk and soul as relating to works which 'simultaneously referenced a past of abduction, displacement and alien-nation [*sic*]' and inspire 'technical and creative innovations'. Here, science fiction becomes a recurring motif within black expressive forms because it is 'an apt metaphor for black life and history' (ibid.). Eshun (1998), for example, traces an Afrofuturist strain across jazz, funk, techno, hip-hop and British bass music to argue that posthumanity and alienness are key articulations in black subjectivity. For Eshun, the dehumanizing thrust of slavery is negotiated within certain forms of black popular culture through circumventing the modality of the human towards a form of posthumanism that is both critical and utopian. In a direct engagement with Afrofuturist perspectives, Weheliye (2002) identifies an emergent thematic preoccupation with contemporary technology within R&B and hip-hop in which the post-human voice (as articulated through what he calls the telephone and vocoder effects) becomes a kind of 'sonic "cinema verité" that depict[s] the "reality" of current technologically mediated life worlds' (2002, 33). So instead of being a direct expression of an Afrofuturist anti-humanist tradition or necessarily a reflection of 'the alien', the 'increased prominence of ... technological artifacts in R&B indicates the enculturation (the ways in which technological artifacts are incorporated into the quotidian) of informational technologies in cultural practices' (ibid.). Furthermore, Weheliye positions this trend as being a reconfiguration of desire within R&B towards non-human objects which moves 'desire from the realm of the ideal to the crassly material' (2002, 39). The use of Auto-Tune in these contexts feeds into and reflects the highly processed aspirational aesthetic of mainstream pop.

This is not to say that the technologically mediated voices are not positioned in relationship to a black music continuum. Often the use of Auto-Tune in R&B and hip-hop is less about digital perfection in pitch than a highlighting of, and engagement with, identifiably 'black' vocal traits. The extreme use of Auto-Tune is often employed to work against conventions in vocal delivery. In the digitized melisma, Auto-Tune works to clash against vocal flourishes in order to create a kind of digital growl. The extreme quantization creates a glitchy quality that has become a key generic signifier within hip-hop and R&B. The use of the technique provides a heightened and self-conscious technological mediation of the qualities of soulfulness as inherited from the formal and aesthetic conventions of the expressive continuum of the black diaspora. As Weheliye (2002) goes on to note, the technological framing of vocals within 'black popular music amplifies the human provenances of the voice, highlighting its virtual embodiment, because it conjures a previous, and allegedly more innocent, period in popular music, bolstering the "soulfulness" of the human voice'. Auto-Tune thus both echoes and remakes black music's Afrofuturist tradition through a fit between technological effect and stylistic tradition. The importance of the connection between the audible effects of Auto-Tune and formal musical attributes of the diasporic tradition is primary. This is borne out by the fact that the effect has been adopted enthusiastically by producers and artists working in other melismatic traditions that lack an overtly futuristic discourse. This can be heard most prominently in North African Arabic singing styles. From the Algerian Rai of Chaba Djenet ('Kwit Galbi Wahdi' 2000) to the Tuareg guitar music of Mdou Moctar ('Anar' 2013), Auto-Tune can be heard across genres, often on recordings which have otherwise have fairly traditional, organic production aesthetic. As Clayton (2009) notes:

> Melisma is equally if not more prevalent in Maghrebi music [than in the African American continuum]. This explains the plug-in's mind-boggling success across North Africa. Contemporary raï and Berber music embrace Auto-Tune so heartily precisely because glissandos are a central part of vocal performance (you can't be a good singer unless your voice can flutter around those notes): sliding pitches sound startling through it.

Nevertheless, it is clear that there has been something of a stylistic conjugality between the qualities of the heavy use of Auto-Tune and vocal articulations and characteristics which have been culturally located as 'black'. As a result the effect has become a cultural signifier that has come to have a relatively wide emotional and expressive range despite its perceived tendency to homogenize vocal timbre. For example, Kanye West's 2008 album *808s and Heartbreak* uses the technique over the course of a melancholic and introspective set of songs. Throughout the album's mix of downbeat electropop and R&B style ballads, West's vocal address has a sense of intimacy and sensitivity in which Auto-Tune is used to frame and accentuate rather than mask the expressive and emotional signifiers of the voice. Even on the album's most explicit lyrical reference to cyberculture, his heavily Auto-Tuned voice berates an ex-lover for a coldness and lack of humanity, asking 'when did you become a Robocop?' In the context of the song there is no inconsistency between the fact that the human, feeling subject position of the protagonist is articulated through a clearly electronically manipulated voice. The limitation in frequency range on the overall voice produced by the effect is used to construct a plaintive and even vulnerable mode of vocal address. This is in contrast to other performances of masculinity via Auto-Tune in hip-hop such as Lil' Wayne's 'Lollipop' and Snoop Dogg's 'Sensual Seduction' whereby the effect is used to create a more detached address in which 'sexual desire [is made to] sound automated and automatic, reducing sex to mechanics rather than a play of emotions' (Rogers 2009, 48). In fact, West's use of Auto-Tune in this manner is indicative of the diversity of emotional and discursive signification. As Ledsham (2015) argues, it is tempting to see the ubiquity of the effect as instrumental in 'a broadening of the often narrow emotional colours previously open to rappers'. A survey of 2015 R&B and hip-hop reveals the continuing ubiquity of the effect and a multiplicity in its applications from the sensual, sexualized female address of DeJ Loaf's 'Me U & Hennessy', the defiant and aspirational tone of Rae Sremmurd's 'This Could Be Us', Young Thug Feat Duke's attempt to audibly replicate the effects of misused prescription drugs ('With That') to the smooth, almost crooner-like masculinity of Jidenna's 'Classic Man'.

The diversity in the application of explicit Auto-Tune use thus has much to tell us about the cultural framing of technological–human interactions. The enculturation of Auto-Tune and the multitude of differing uses across the contemporary soundworld of popular music illustrated by the historical trajectory outlined here are indicative of the pervasiveness and transparency of technologies to contemporary post-human subjectivities. The fact that Auto-Tune's meanings are multiple and unfixed suggests that for producers and audiences there is little contradiction between the technological staging of the voice and its use to convey a variety of differing human emotions. Such a normalization and naturalization of the technologically mediated voice questions the binaries between the organic and the mechanic that critical posthumanist theory has sought to problematize (Nayar 2013, 19). As Clayton notes, the multiple and often contradictory use of Auto-Tune is indicative of how developed and intimate our relationship with technology has become:

> The plug-in creates a different relation of voice to machine than ever before. Rather than novelty or some warped mimetic response to computers, Auto-Tune is a contemporary strategy for intimacy with the digital. As such, it becomes quite humanizing. Auto-Tune operates as a duet between the electronics and the personal. (Clayton 2009)

Intrinsic use of Auto-Tune

On a more insidious level, Auto-Tune's ubiquity in the modern studio set-up has also contributed to a broader aesthetic of digital perfectionism that permeates contemporary pop. One the one hand, this has resulted in a (technologically mediated) consistency in pitching in contemporary pop recordings, and on the other, a subtle shift in conventions of vocal timbre. The intrinsic use of Auto-Tune is hidden perhaps due to public perceptions that its use somehow denigrates the authenticity of a performance (Hughes 2015). Lady Gaga is one of the few mainstream performers to admit to extensively using intrinsic Auto-Tune. Even in this instance her comments were in the context of a career repositioning through a duet project with the veteran crooner Tony Bennett and drew upon well-worn debates around authenticity and control. The thrust of her comments was that her record company had insisted on the use of Auto-Tune in order to position her material within the contemporary production values demanded by the marketplace. She commented that all throughout her career 'they've been auto-tuning it more … changing the timbre. They take the vibrato out so you sound like a robot. … Although it was still my songs, and I still had a lot to say about the production, the vocal was something that they really, really wanted to control' (McLean 2014).

The fact that the technique is to a certain extent shrouded in secrecy is perhaps complicit in its naturalization. Auto-Tune is ubiquitous yet unannounced and actively hidden, leading to a normalization of the sonic qualities it produces within public understandings of what a pop singing voice *should* sound like. While intrinsic usage of the effect is designed to be seemingly imperceptible, it clearly lends particular characteristics to the voice leading to a pervasive but less explicit vocal production style on modern recordings. This is largely to do with how pitch correction has an effect upon the overall tonal qualities of the voice. In addition to the unnaturally fast transition between notes without glissando, the use of Auto-Tune has other perceptible effects. Generally, there is something of an inability to capture the full resonant range of the voice in terms of natural harmonics and levelling out of the micro nuances of vibrato, resulting in a slightly synthetic tonal modulation. The singing voice is made up of a complex range of frequencies that contribute to timbre and fundamentally shape our understanding of the individuality of a given voice. As well as being made up of a fundamental pitch, the voice also produces overtones, that is, secondary pitches or formants which correspond to multiples of the fundamental frequency. It is the fact that the fundamental is not heard in isolation that contributes to our understanding of a voice as 'rich' or 'full'. The intrinsic use of Auto-Tune has the tendency to flatten out some of these overtones resulting in an even vocal timbre, which, while being less explicitly technologically staged than in its explicit uses, has nonetheless become an omnipresent and definable feature in the vocal staging of contemporary pop music. While the results of this processing may be subtle, they are clearly present. This is borne out by Keith's (2014, 8) spectrographic visualizations of Auto-Tuned vocal tracks which illustrate a removal of microtonal variations resulting in a 'stable central [harmonic] frequency with minimal variation', a dynamic flatness in which 'the voice's energy remains constant' and a level of 'formant variation' in which partial harmonics are lost. The resultant sound is therefore not only a digital manipulation of pitch but also elicits a digitally constructed consistency in timbre and dynamics that permeates contemporary popular music, all of which contribute to the construction of contemporary ideals of vocal perfection.

These characteristics provide a subtle shift in the staging of the contemporary pop voice and permeate the landscape of production so widely as to be transparent. This naturalization can also be understood through how the sonic side effects of Auto-Tune intersect and chime with the digital soundworld of modern productions. There is a clear aesthetic fit between Auto-Tune as vocal staging and a production style which is based upon the extensive use of synthesizers, digital reverbs and delays and an overall tight tonal range accentuated by the extensive use of compression and the foregrounding of a single layered vocals within the mix. Take, for example, Jess Glynne's 2015 international hit 'Hold My Hand' where, although being employed throughout the track, Auto-Tune is most discernable in the breakdown section (at 2 minutes 45 seconds into the track) where it is exposed in the context of the stripped down musical accompaniment in the section. Here, there is a clearly perceptible flattening of some of the natural tonal variation in the voice and a subtle but still audible erosion of glissando between notes typical of modern pop production. Rather than sounding conspicuous within this context, the effect on vocal timbre here is highly congruent with other aspects of production (obviously digital piano and horn sounds, electronic drum beats, filtered delays on the vocal, etc.). There are clear parallels here with other forms of vocal staging. For example, while Lacasse (2000) is interested in the spatial semiotics of the voice in recordings, he nevertheless points to the use of reverb on the voice as being a naturalized and almost expected framing device within popular music recordings. Similarly, Auto-Tune's repeated use and the subsequent sonic qualities lent to recorded texts have become equally naturalized within the listening expectations of contemporary pop audiences leading to perfect pitching and its timbral inflections becoming almost expected within certain types of popular music texts. For example, in a discussion of his common production practices, Carlo 'Illangelo' Montagnese who has worked with globally successful pop acts such as the Weeknd, Drake, Lady Gaga, MIA, and Florence and the Machine indicated that he always uses Auto-Tune in his vocal productions even if there were no apparent pitching issues with a singer's take. He commented: 'If you don't pull a vocal through Auto-Tune these days, it almost does not sound normal! So you have to use it, though I apply it very subtly. You might not even notice it, but for me it makes a huge difference' (Tingen 2015).

On the one hand, this perhaps leads to particular expectations in audiences in terms of pitching within singing voices more generally and, on the other, to what is considered normal or natural in the timbre of the recorded voice. Hence, even in intrinsic uses that are not immediately aurally perceptible, Auto-Tune's almost total ubiquity in contemporary vocal staging has implications for common conceptions of the idealized vocal performance and ultimately the qualities of the human voice. For example, there is also some evidence to suggest that the technology is having an effect upon singing practice itself. In a discussion of contemporary issues of singing pedagogy for instance, Frazier-Neely (2013, 594) sees young singers' habits in terms of stylistic aspirations and expectations along with the entrainment of the body in relationship to the voice as shifting as they are constantly exposed to conventions mediated by technological manipulations of the voice.[6] While Zagorksi-Thomas

[6]For example, in August 2015 a YouTube video of 18-year-old amateur singer Emma Robinson covering Tori Kelly's hit 'Paper Hearts' went viral, amassing over two million hits after viewers noted that her (un-Auto-Tuned) tone and use of melisma bore the characteristics of the pitch correction technology. See also a discussion thread on the producer's forum Gearslutz.com where a number of producers comment on commonly encountering young singers who had taken on an emulation of the stylistic and timbral attributes of Auto-Tune in their singing voices. https://www.gearslutz.com/board/so-much-gear-so-little-time/393226-vocalists-singing-like-Auto-Tune-without-Auto-Tune.html (accessed 20th February 2016).

(2014, 64) points out that notions of perfection in recording convention are socially constructed, there is no doubt that the structure of idealized notions of the voice here is inherently socio-technological. The plug-in produces a technologically mediated articulation of perfection through a digitally framed construction of what is the 'right' note. As Hajdu (2012) notes, 'It applies ... [a] rigid definition of rightness' by adjusting 'every tone with unyielding, unvarying precision, squarely in the mathematical center of the note.'

This digitally constructed aesthetic of perfection is indicative of the wider cultural implications of digitization. The vocal staging of contemporary pop chimes with the constructed and aspirational notion of perfection that permeates many aspects of contemporary culture more generally. Current modes of visual and auditory culture within advertising photography, film, television all use digital technologies to construct a hyperreal mediated version of reality which conforms to certain ideals of beauty, aesthetics and desirability. Technologies such as Photoshop in photography and Digital Intermediate in film and television, through their repeated use and their integration as industry-standard techniques, construct digitally manipulated versions of the world that have become dominant. The construction of a digital perfection in contemporary pop music should thus be understood as a utilization of contemporary technologies and as part of a wider landscape of dominant cultural representation. In turn, we can relate the digital construction of perfection to considerations in post-human subjectivities. Our interactions and utilizations of digital technologies in our daily lives actively engender the editing, construction and presentation of a sense of ourselves that is digitally mediated. Nowhere is this more apparent than in the integration of social networking into our social interactions and identities. As McNeil (2012, 71–2) argues, the particular structures of interaction and self-representation imposed by the functionality of sites such as Facebook have a material effect upon how we structure and understand a sense of autobiography and narrative in our lives in a 'post-human process of identity formation'. She notes that the logics of the site encourage self-surveillance, self-editing and the construction of exemplary lives and behaviours through technology (2012, 74). This assessment I think, belies a more pervasive consequence of human–computer relationships in the digital age. In a sense, if we engage with digital technologies we are engaged in processes of editing, manipulation and enhancement. The enculturation of the positioning of the human voice within technological mediation is therefore reflective of our wider relationship to technology and is a logical expression of the material conditions of digitalization.

Summary

The sociotechnical situatedness of contemporary popular music aesthetics and creativity within the material conditions of digitization are key to understanding the examples covered in this chapter. The emergence of the various articulations of digital aesthetics discussed here is simultaneously reflective of changes in how music is composed, recorded and produced as a result of digital technologies and a wider integration of digital technologies in our daily lives. First, the way in which the working environment of the DAW engenders particular workflows, specific types of creative trajectories and sonic characteristics has had a fundamental effect upon the sounds and structures of contemporary popular music across a variety of genres. For example, glitch, hauntology and vaporwave constitute very different articulations of the materialist process facilitated by DAW technologies that has become central to creativity within postmillennial genres. While each of the examples outlined in the first half of the chapter are resultant from divergent musical cultures and histories, they all share an embedded naturalization of digital technologies in their common creative strategies and mediation. They are exemplary of the level of enculturation of DAW technologies across the spread of musical cultures in the contemporary creative environment. Similarly, the widespread use of pitch correction technologies on the voice across a variety of textual sites from mainstream pop to a variety of African American and African genres is indicative of how the functionality of VST plug-ins has fundamentally altered conventions of vocal staging and ideas of vocal perfection in a way which traverses generic and geographical boundaries.

Secondly, the adoption and acceptance of these strategies into a broad aesthetic field also tells us something about wider computer–human relationships in the context of digitization. The ways in which digital technologies pervade our subjectivities and traverse our spaces of work, leisure and identity have provided a social and cultural context in which digital aesthetics *make sense*. As listeners, we require music to have a social use and value that is reflective of our lived realities and pertinent to our needs. We negotiate and construct our identities through music, we punctuate our daily lives through its use; we entrain our minds and bodies in moments of relaxation, arousal and concentration through listening. Given our increasing reliance upon and comfort with digital technologies, our day-to-day exposure to digitally constructed and manipulated texts within cultural interactions and consumption, it is consistent that these functions of music are unproblematically played out through an overtly digital soundscape. The variety of digital aesthetics outlined throughout this chapter, both in their emergence and reception, are inextricably linked to the lived realities of digitization of the past two decades.

Conclusion

There can be little doubt … that music is an indicator of the age, revealing for those who know how to read its symptomatic messages, a means of fixing social and even political events.

(SHAFER 1977, 7)

On a fundamental level the arguments presented within this book should be understood as being engaged with the ongoing entwinement of popular music and technology. Indeed, the capacities and developments of technologies are so central to music per se that any convincing account of its developmental history should also simultaneously be an account of technology. This is the case with any genre of music (e.g. the importance of technological development in Western classical music since the invention of musical instruments). While the specific focus here has been upon electronic/digital technology, 'technology' itself involves any form of artefact design, construction and use. In turn, the development of these artefacts clearly has material effects upon practice. The relationship between, say, Bach and the equal temperament, Bach/Mozart/Beethoven and the passage from clavichord to pianoforte to grand piano (which of course had a direct effect on the creative process and the musical pieces themselves) or the development of the violin are all examples of technological/social/aesthetic relationships. Yet, there has been something of a resistance from within academe to consider Western art music in these terms, perhaps driven by an ideological drive to preserve classical music from a materialist critique in order to preserve its place within in a 'humanist' perspective. This is something of a missed opportunity. The types of insights offered by social theories such as ANT and the production of culture perspective that have been central to this book are not just applicable to 'new' technologies but may equally offer insights in terms of historical musicology of all persuasions.

Perhaps the centrality of the mediation of popular music through sound recording (along with its long history of interdisciplinarity) has meant that popular music studies scholarship has more naturally gravitated towards such materialist approaches. Significant work from this perspective has shown how from the advent of recorded music in the nineteenth century recording and playback technologies have had fundamental effects upon how music sounds, what it means and how it is understood. From the bodily entrainment of the human voice in the days of acoustic recording (Sterne 2006) through to the invention of the microphone (Frith 2007) to the development of multitrack and tape-recording (Zak 2001), advances in technologies have been seen to be instrumental in paradigm shifts relating to what musicians do and how they think about the musical object. Similarly, a parallel body of work has demonstrated how each shift in playback technology from the gramophone (Katz 2004) to the MP3 player (Sterne 2006a; Bull 2007) has brought with it a set of structural and aesthetic changes that have had a profound effect upon popular music culture.

The account of recent developments outlined throughout this book constitutes an attempt to contribute to our understanding of digital technology's central place within this multiplicity of historical trajectories. More specifically, it has highlighted a number of interconnected developments resultant from a rapid period of change from the 1990s ushered in by convergent digitization. As such, this book takes Schafer's evocation of music as a revelator of the historical specificities of social realities as self-evident. A core implication is that significant changes in popular music practice and culture are inherently situated in, and reflective of, wider patterns of cultural and technological change. This book has examined the changes that have occurred in the digital landscape and the profound effects that they have had upon the sonic characteristics of music, its production practices, its consumption and the blurring of distinctions between professional and amateur practitioners.

The processes through which these changes have happened are complex and interrelated but there are a number of key implications we can draw in terms of how music is created, who has access to the means of production and how that music is heard and shared. Perhaps the most important factor has been the increasing centrality of the personal computer as a site of creativity across a broad range of musical practice. This has led to a growing number of people engaging with computer technology as a way of making music and to a restructuring of how musical creativity is conceptualized resultant from a reliance on the digital representation of sonic material and the realization of compositional/production tasks through the visual metaphors of personal computing. In turn, these shifts have been instrumental in the escalating ubiquity of digitally produced electronic sound within the day-to-day soundscapes of popular music. Finally, these transformations have unfolded within, and have contributed to, a cultural and industrial context that has seen dramatic shifts in the dominant modes of music distribution and consumption.

This coalescence of technology and creativity can be seen to reflect a posthumanist/transhumanist vision of artistic practice founded on the use of digital artefacts, to the point that technology might be understood as an extension of the human. In addition, dominant trends in the musics produced through technical/human/social coalitions provide an 'indicator of the age' and do indeed reveal 'symptomatic messages' about how we live and interact in the post-human context. The enculturation and centrality of digitally produced and manipulated sound as a form of human expressivity within popular music belies a tacit acceptance of the material conditions of digitalization. In this sense, the practices of creativity and development of aesthetics outlined within the proceeding discussion are part of wider post-digital shifts whereby our experience, subjectivities and actions are all enmeshed in the context of an advanced stage of HCIs. They are symptomatic of how the many divergent threads of our lives are increasingly converging upon multifunctional, multipurpose digital technologies and the ways in which our thinking and acting are mediated by this interaction.

The differing theoretical strands brought together in this book have all pointed to similar movements in our relationship to digital technologies in the contemporary cultural context. Although ANT and affordance theory were not primarily concerned with digital technologies in their original incarnations, they have much to offer in terms of explaining the nature of our increasing reliance upon digital artefacts. The idea that humans are actants in techno-social assemblages or that our actions should be considered in terms of their place within an ecology of objects and animals has never been so pertinent. A conceptual repositioning of the individual in the developed world as post-human 'implies not only a coupling with intelligent machines but a coupling so intense and multifaceted that it is no longer possible to distinguish meaningfully between the biological organism and the informational circuits in which the organism is enmeshed' (Hayles 1999, 35). As Braidotti (2013, 57) posits, in the current context the 'posthuman subject is technologically mediated to an unprecedented degree'. In other words, the way in which we experience the world and interact within it, our patterns of behaviour and our sense of identity have been shaped, and continue to be shaped, by recent technological developments.

The accelerated situation of creativity within the context of a highly technologically mediated and digitized modality of popular music production should be no surprise. Given our increasing reliance on digitally mediated cultural forms, it also may seem self-evident that the grain of our everyday musical experience should be progressively more suffused with the sonic signifiers of the digital. Yet, it is important to critically assess and document these social facts, even if axiomatic. Ultimately, this book's positioning of the personal computer as the dominant 'sonic technology' of our age highlights the need to take account of the specificities of the strains within changing HCIs that shape our cultures. We need to mark out patterns of enculturation, trace the technological scripts that have become 'transparent', map out the terrain of digital culture and listen intently to its emergent soundscapes. We must also, as scholars, take into account new digital formations as they emerge. The analysis presented in this book merely sketches a picture of musical practice and creativity at the start of a post-digital age. Technological development will no doubt result in the further integration of digital artifacts within the creative process; increasing geographical fragmentation of the sites of creativity into cybernetic spaces, and forging new performance conventions and compositional strategies as haptic and virtual technologies become increasingly prevalent. As these nascent technological developments become increasingly more central to our musical worlds it is imperative that we examine them holistically through an approach which takes into account the relationship between the artefact, the individual and social.

WORKS CITED

Adegoke, Y. (2006), 'Long-Tailed Niche Market of the Future Where Less is More', *Marketing Week*, 27 July 2006: 24–5.

Aho, M. (2009), '"Almost Like the Real Thing": How Does the Digital Simulation of Musical Instruments Influence Musicianship', *Music Performance Research*, 3: 22.

Akrich, M., and B. Latour (1992), 'A Summary of a Convenient Vocabulary for the Semiotics of Human and Nonhuman Assemblies', in Bijker and Law (eds), *Shaping Technology/Building Society: Studies in Sociotechnical Change*, 259–64, Cambridge, MA: MIT Press.

Antares (2011), *Auto-Tune EFX Real-Time Auto-Tune Vocal Effect and Pitch Correcting Plug-In Owners Manual*, Scotts Valley: Antares Audio Technologies.

Arditi, D. (2014), 'Digital Downsizing: The Effects of Digital Music Production on Labor', *Journal of Popular Music Studies*, 26(4): 503–20.

Ashline, W. (2002), 'Clicky Aesthetics: Deleuze, Headphonics, and the Minimalist Assemblage', *Strategies: Journal of Theory, Culture, Politics*, 15(1): 87–105.

Augoyard, J. F., and H. Torgue (2006), *Sonic Experience: A Guide to Everyday Sounds*, Montreal: McGill-Queens University Press.

Auslander, P. (1999), *Liveness: Performance in a Mediatized Culture*, London: Routledge.

Bader, I., and A. Scharenberg (2010), 'The Sound of Berlin: Subculture and the Global Music Industry', *International Journal of Urban and Regional Research*, 34(1): 76–91.

Bærentsen, K. B., and Trettvik, J. (2002), 'An Activity Theory Approach to Affordance', in O. W. Bertelsen, S. Bødker and K. Kuuti (eds), *Proceedings of the Second Nordic Conference on Human-Computer Interaction*, Aarhus, Denmark, 51–60.

Barker, C. (2012), 'Alesso Explains the Making of Years', *Futuremusic,* 28 November. http://www.musicradar.com/news/dj/alesso-explains-the-making-of-years-567223/ (accessed 20 January 2015).

Barker, C. (2013), 'Clean Bandit Talk About the Making of Mozart's House', *Futuremusic,* 24 May 2013. http://www.musicradar.com/news/tech/clean-bandit-talk-about-the-making-of-mozarts-house-575391/ (accessed 20 January 2015).

Bates, E. (2004), 'Glitches, Bugs, and Hisses: The Degeneration of Musical Recordings and the Contemporary Musical Work', in C. Washburne and M. Derno (eds), *Bad Music: The Music We Love to Hate,* 275–93, London: Routledge.

Bates, E. (2009), 'Ron's Right Arm: Tactility, Visualization, and the Synesthesia of Audio Engineering', *Journal on the Art of Record Production*, 04. http://arpjournal.com/rons-right-arm-tactility-visualization-and-the-synesthesia-of-audio-engineering/ (accessed 10 May 2016).

Baudrillard J. (1988), 'Simulacra and Simulations', in M. Poster (ed.), *Jean Baudrillard, Selected Writings,* 166–84, Stanford: Stanford University Press.

Becker, H. (1982), *Art Worlds*, Berkeley: University of California Press.

Bell, A. P. (2015), 'DAW Democracy? The Dearth of Diversity in "Playing the Studio"', *Journal of Music, Technology and Education,* 8(2): 129–46.
Bell, A. P., E. Hein and J. Ratcliffe (2015), 'Beyond Skeuomorphism: The Evolution of Music Production Software User Interface Metaphors', *Journal of the Art of Record Production*, 09. http://arpjournal.com/beyond-skeuomorphism-the-evolution-of-music-production-software-user-interface-metaphors-2/ (accessed 05 May 2016).
Bennett, A., B. Shank and J. Toynbee (2006), 'Introduction', in A. Bennett, B. Shank and J. Toynbee (eds), *The Popular Music Studies Reader*, 1–9, Routledge: Abingdon.
Benson, C. (1993), *The Absorbed Self: Pragmatism, Psychology and Aesthetic Experience*, London: Harvester Wheatsheaf.
Bertelsen, O. V. (2006), 'Tertiary Artifacts at the Interface', in P. Fishwick (ed.), *Aesthetic Computing,* 357–68, Cambridge, MA: MIT Press.
Biddle, I. (2004), 'Vox Electronica: Nostalgia, Irony and Cyborgian Vocalities in Kraftwerk's Radioaktivität and Autobahn', *Twentieth Century Music*, 1: 81–100.
Blackwell, A. F. (2006), 'The Reification of Metaphor as a Design Tool', *ACM Transactions on Computer-Human Interaction*, 13(4): 490–530.
Bolter J., and D. Gromala (2003), *Windows and Mirrors: Interaction Design, Digital Art and the Myth of Transparency*, Cambridge, MA: MIT Press.
Bolter J., and D. Gromala (2006), Transparency and Reflectivity: Digital Art and the Aesthetics of Interface Design', in P. Fishwick (ed.), *Aesthetic Computing*, 369–82, Cambridge, MA: MIT Press.
Bolter, J., and Grusin, R. (2000), *Remediation: Understanding New Media*, Cambridge, MA: MIT Press.
Borghi, A. M., and L. Riggio (2009), 'Sentence Comprehension and Simulation of Object Temporary, Canonical and Stable Affordances', *Brain Research, 9*: 117–28.
Born, G. (1995), *Rationalizing Culture: IRCAM, Boulez, and the Institutionalization of the Musical Avant-Garde*, Berkeley: University of California Press.
Bostrom, N., and R. Roache (2007), 'Ethical Issues in Human Enhancement', in J. Ryberg, T. S. Petersen and C. Wolf (eds), *New Waves in Applied Ethics*, 120–52, London: Palgrave-Macmillan.
Bourdieu, P. (1971), 'Intellectual Field and Creative Project', in M. F. D. Young (ed.), *Knowledge and Control: New Directions in the Sociology of Education,* 161–88, London: Collier-Macmillan.
Bourdieu, P. (1984), *Distinction: A Social Critique of the Judgement of Taste*, Boston: Harvard University Press.
Bourdieu, P. (1993), *The Field of Cultural Production*, Cambridge: Polity.
Bouwman, K. (2009), 'Interview with Mich "Cutfather" Hansen', 2 November 2009, *Hit Quarters*. http://www.hitquarters.com/index.php3?page=intrview/opar/intrview_Mich_Hansen_Interview.html (accessed 21 November 2015).
Braidotti, R. (2013), *The Posthuman*, London: Polity.
Brennan, T. (2004), *The Transmission of Affect*, Ithaca, NY: Cornell University Press.
Brennen, S., and D. Kreiss (2014), 'Digitalization and Digitization', *Culture Digitally*, 8 September 2014. http://culturedigitally.org/2014/09/digitalization-and-digitization/#sthash.WdAGrxpy.dpuf (accessed 1 February 2016).

Brøvig-Hanssen, R. (2013), 'Opaque Mediation: The Cut-And-Paste Groove in DJ Food's "Break"', in A. Danielsen (ed.), *Musical Rhythm in the Age of Digital Reproduction*, 159–76, Farnham: Ashgate.

Brøvig-Hanssen, R., and A. Danielsen (2016), *Digital Signatures: The Impact of Digitization on Popular Music Sound*, Cambridge, MA: MIT Press.

Bruns, A. (2010), 'Distributed Creativity: Filesharing and Produsage', in Stefan Sonvilla-Weiss (ed.), *Mashup Culture*, 24–37. New York: Springer.

Buckingham, D., and R. Willet (2009), *Video Cultures Media Technology and Everyday Creativity*, Basingstoke: Palgrave.

Bull, M. (2007), *Sound Moves: IPod Culture and Urban Experience*, London: Routledge.

Campbell-Kelly, M. (2003), *From Airline Reservations to Sonic the Hedgehog: A History of the Software Industry*, Cambridge, MA: MIT Press.

Cascone, K. (2000), 'The Aesthetics of Failure: "Post-digital" Tendencies in Contemporary Computer Music', *Computer Music Journal*, 24(4): 12–18.

Challacombe, C., and E. Block (2014), *The 2014 NAMM Global Report*, Carlsbad, CA: NAMM.

Chambers, I. (1992), 'Contamination, Coincidence and Collusion: Pop Music, Urban Culture and the Avant-Garde', in P. Brooker (ed.), *Modernism/Postmodemism*, 190–6, London: Longman.

Chesher, C. (2004), 'Neither Gaze Nor Glance, But Glaze: Relating to Console Game Screens', *SCAN Journal of Media Arts Culture*, 1(1). http://scan.net.au/scan/journal/display.php?journal_id=19 (accessed 14 January 2012).

Chinn, K. (1985), 'Atari Announces Six New Computers', *InfoWorld*, 28 January 1985: 15–16.

Chopik, I. (2010), 'Misha Mansoor Interview (Periphery)', *Guitar Messenger*, 24 April 2010. http://www.guitarmessenger.com/interviews/misha-mansoor-interview-periphery/ (accessed 5 February 2016).

Clarke, E. F. (2003), 'Music and Psychology', in M. Clayton, T. Herbert and R. Middleton (eds), *The Cultural Study of Music: A Critical Introduction*, 113–23, London: Routledge.

Clarke, E. F. (2005), *Ways of Listening: An Ecological Approach to the Perception of Musical Meaning*, Oxford: Oxford University Press.

Clayton, J. (2009), 'Pitch Perfect: The Highs and Lows of "Auto-Tune", the Software used in Almost Every Pop Song Released Today', *Frieze* Issue, 129. http://www.frieze.com/issue/article/pitch_perfect/ (accessed 15 November 2015).

Cohen, I. (2014), 'Update: Girl Talk', *Pitchfork*, 25 March 2014. http://pitchfork.com/features/update/9362-girl-talk/ (accessed 1 August 2015).

Collins, N., and J. d'Escriván (2009), *The Cambridge Companion to Electronic Music*, Cambridge: Cambridge University Press.

Cone, E. (1968), *Musical Form and Musical Performance*, New York: Norton.

Cooke, C. (2015), 'Indies Enter the Breakage Debate', *Complete Music Update*, Monday, 8 June 2015. http://www.completemusicupdate.com/article/indies-enter-the-breakage-debate/#sthash.qxA7oZo3.dpuf (accessed 10 August 2015).

Csikszentmihalyi, M. (1988a), *Optimal Experience: Psychological Studies of Flow in Consciousness*, Cambridge: Cambridge University Press.

Csikszentmihalyi, M. (1988b), 'Society, Culture and Person: A Systems View of Creativity', in R. Sternberg (ed.), *The Nature of Creativity: Contemporary Psychological Perspectives*, 325–39, New York: Cambridge University Press.

Csikszentmihalyi, M. (2008), *Flow: The Psychology of Optimal Experience*, New York: Harper.
Cypher, M., and I. Richardson (2006), 'An Actor-Network Approach to Games and Virtual Environments', *Joint Computer Games and Interactive Entertainment Conference*, Perth Australia.
Daley, D. (1999), 'Recordin' La Vida Loca: The Making of a Hard-Disk Hit', *Mix*, 11 January 1999. http://www.mixonline.com/news/profiles/recordin-la-vida-loca-making-hard-disk-hit/374667 (accessed 20 January 2015).
Dax, M. (2013), 'Self-Liberating: An Interview with Grimes', *Electronic Beats*, 03 January 2013. http://www.electronicbeats.net/self-liberating-an-interview-with-grimes/ (accessed 02 March 2015).
Décary-Hétu, D. (2014), 'Police Operations 3.0: On the Impact and Policy Implications of Police Operations on the Warez Scene', *Policy and Internet*, 6(3): 315–40.
Deleuze, G. (1986), *Cinema 1: The Movement Image*, Minneapolis: University of Minnesota Press.
Deleuze, G., and F. Guattari (1983), *Anti-Oedipus: Capitalism and Schizophrenia*, Minneapolis: University of Minnesota Press.
Deleuze, G., and F. Guattari (1994), *What Is Philosophy*, London: Verso.
Demers, J. (2010), *Listening Through the Noise: The Aesthetics of Experimental Electronic Music*, Oxford: Oxford University Press.
DeNora, T. (2000), *Music in Everyday Life*, Cambridge: Cambridge University Press.
DeNora, T. (2003), *After Adorno: Rethinking Music Sociology*, Cambridge: Cambridge University Press.
Dewey, J. (1980), *Art as Experience*, New York: Perigee.
Dickinson, K. (2001), 'Believe? Vocoders, Digitalised Female Identity and Camp', Source: *Popular Music*, 20(3): 333–47.
Ding, W., and X. Lin (2010), *Information Architecture: The Design and Integration of Information Spacesm*, San Francisco: Morgan and Claypool.
Doyle, P. (2005), *Echo and Reverb: Fabricating Space in Popular Music Recording, 1900–1960*, Newtown, CT: Wesleyan University Press.
Du Gay, P. (1997), *Doing Cultural Studies: The Story of the Sony Walkman*, Milton Keynes: Oxford University Press.
Duignan, M., J. Noble and R. Biddle (2010), 'Abstraction and Activity in Computer-Mediated Music Production', *Computer Music Journal*, 34(4): 22–33.
Dunsby, J. (1983), 'Music and Semiotics: The Nattiez Phase', *The Musical Quarterly*, 69(1): 27–43.
Economomist (2009), 'A World of Hits', *Economist*, November 2009. http://www.economist.com/node/14959982 (accessed 1 February 2016).
Ellis, J. (1982), *Visible Fictions*, London: Routledge.
Emmerson, S. (2007), *Living Electronic Music*, Aldershot: Ashgate.
Eshun, K. (1998), *More Brilliant Than the Sun: Adventures in Sonic Fiction*, London: Quartet Books.
Ezra, E. E. (2014), 'Posthuman Memory and the Re(f)use Economy', *French Cultural Studies*, 25(3–4): 378–86.
Fabbri, F. (1982), 'A Theory of Musical Genres: Two Applications', *Popular Music Perspectives*, 1: 52–81.
Fairchild, C. (2008), *Pop Idols and Pirates: Mechanisms of Consumption and the Global Circulation of Popular Music*, Aldershot: Ashgate.
Feifer, J. (2012), 'We Are All T-Pain', *Fast Company*, January 2012, 36–8.

Fikentscher, K. (2000), '"*You Better Work!" Underground Dance Music in New York'*, Newtown, CT: Wesleyan University Press.
Fink, R. (2005), 'The Story of ORCH5 or the Classical Ghost in the Hip-Hop Machine', *Popular Music*, 24(3): 1–17.
Fintoni, L. (2015), 'How Fruity Loops Changed Music-Making Forever', *Red Bull Music Acadamy Daily*, 13 May 2015. http://daily.redbullmusicacademy.com/2015/05/fruity-loops-feature (accessed 11 November 2015).
Frazier-Neely, C. (2013), 'Live vs. Recorded: Comparing Apples to Oranges to Get Fruit Salad', *Journal of Singing*, 69(5): 593–6.
Freidman, T. (2005), *Electric Dreams: Computers in American Culture*, New York: New York University Press.
Frith, S. (2007), *Taking Popular Music Seriously*, Aldershot: Ashgate.
Frith, S. (2011), 'Creativity As A Social Fact', in D. Hargreaves, D. Miell and R. MacDonald (eds), *Musical Imaginations: Multidisciplinary Perspectives on Creativity, Performance and Perception*, 62–72, Oxford: Oxford University Press.
Gall, M., and N. Breeze (2005), 'Music Composition Lessons: The Multimodal Affordances of Technology', *Educational Review*, 57(4): 415–31.
Gaver, W. (1993), 'What in the World Do We Hear?: An Ecological Approach to Auditory Event Perception', *Ecological Psychology*, 5: 1–29.
Gerhardt, W. (2008), *Prosumers: A New Growth Opportunity*, San Jose, CA: Cisco Internet Business Solutions Group.
Gibson, J. J. (1966), *The Senses Considered as Perceptual Systems*, Boston: Houghton Mifflin.
Gibson, J. J. (1979), *The Ecological Approach to Visual Perception*, Hillsdale: Lawrence Erlbaum Associates.
Gilbert, J., and E. Pearson (1999), *Discographies: Dance Music, Culture, and the Politics of Sound*, London: Routledge.
Goodwin, A. (1990), 'Sample and Hold: Pop Music in the Digital Age of Reproduction', in S. Frith and A. Goodwin (eds), *On Record: Rock, Pop and the Written Word*, 259–73, New York: Pantheon.
Goodwin, A. (1992), 'Rationalization and Democratization in the New Technologies of Popular Music', in J. Lull (ed.), *Popular Music and Communication*, 75–100, Newbury Park, CA: Sage.
Gracyk, T. (1996), *Rhythm and Noise: An Aesthetics of Rock*, London: Tauris.
Grossman, L. (2006), 'You, Yes You are TIME's Person of the Year', *Time*, 25 December 2006. http://content.time.com/time/magazine/article/0,9171,1570810,00.html (accessed 20 February 2016).
Gunelius, S. (2010), 'The Shift from CONsumers to PROsumers', *Forbes*, 7 March 2010. http://www.forbes.com/sites/work-in-progress/2010/07/03/the-shift-from-consumers-to-prosumers/#60228362543f (accessed 2 February 2016).
Hainge, G. (2007), 'Of Glitch and Men: The Place of the Human in the Successful Integration of Failure and Noise in the Digital Realm', *Communication Theory*, 17: 26–42.
Hajdu, D. (2012), 'Imperfect Pitch', *New Republic*, 22 June 2012. https://newrepublic.com/article/104194/david-hajdu-music-imperfect-pitch (accessed 9 January 2016).
Hampton, J. (2012), 'Skream and Benga: Foundational Figures', *DJ Times*, 1 January 2012. http://djtimes.com/skream-benga-foundational-figures/ (accessed 2 December 2015).

Hancox, D. (2012), 'A History of Grime, By the People Who Created It', *The Guardian,* 6 December 2012. http://www.theguardian.com/music/2012/dec/06/a-history-of-grime.
Harkins, P. (2010), 'Microsampling: From Akufen's Microhouse to Todd Edwards and the Sound of UK Garage', in A. Danielsen (ed.), *Musical Rhythm in the Age of Digital Reproduction*, 177–94, Farnham: Ashgate.
Harkins, P. (2015), 'Following The Instruments, Designers, And Users: The Case Of The Fairlight Cmi', *Art of Record Production Journal*, 10. http://arpjournal.com/following-the-instruments-designers-and-users-the-case-of-the-fairlight-cmi/ (accessed 1 January 2016).
Hawkins, S. (2004), 'Feel the Beat Come Down: House Music as Rhetoric', in A. Moore (ed.), *Analyzing Popular Music*, 80–102, Oxford: Oxford University Press.
Hayles, N. K. (1999), *How We Became Posthuman: Virtual Bodies in Cybernetics, Literature and Informatics*, Chicago: University of Chicago Press.
Hemmungs Wirtén, E. (2004), *No Trespassing: Authorship, Intellectual Property Rights, and the Boundaries of Globalization*, Toronto: University of Toronto Press.
Hesmondhalgh, D. (1996), 'Flexibility, Post-Fordism and the Music Industries', *Media Culture and Society*, 18(3): 469–88.
Hesmondhalgh, D. (1997), 'Post-Punk's Attempt to Democratise the Music Industry: The Success and Failure of Rough Trade', *Popular Music*, 16(3): 255–74.
Hesmondhalgh, D. (1998), 'The British Dance Music Industry: A Case Study in Independent Cultural Production', *The British Journal of Sociology,* 49 (2): 234–51.
Hesmondhalgh, D. (1999), 'Indie: The Institutional Politics and Aesthetics of a Popular Music Genre', *Cultural Studies*, 13 (1): 34–61.
Hesmondhalgh, D. (2007), *The Cultural Industries*, London: Sage.
Hesmondhalgh, D., and S. Baker (2010), '"A Very Complicated Version of Freedom": Conditions and Experiences of Creative Labour in Three Cultural Industries', *Poetics*, 38: 4–20.
Hibbett, R. (2005), 'What is Indie Rock?' *Popular Music and Society*, 28(1): 55–77.
Hofer, S. (2006), 'I Am They: Technological Mediation, Shifting Conceptions of Identity and Techno Music', *Convergence,* 12(3): 307–24.
Holmes, T. (2002), *Electronic and Experimental Music*, London: Routledge.
Houghton, B. (2015), 'In 6 Months Soundcloud Has Approved Just 100 Premier Partners, Payments Total $1 Million', *Hypebot.com,* 3 February 2015. http://www.hypebot.com/hypebot/2015/03/in-6-months-soundcloud-has-approved-just-100-premier-partners-and-paid-them-1-million.html (accessed 20 February 2016).
Hracs, B. (2012), 'A Creative Industry in Transition: The Rise of Digitally Driven Independent Music Production', *Growth and Change,* 43(3): 442–61.
Hughes, D. (2015), 'Technological Pitch Correction: Controversy, Contexts, and Considerations', *Journal of Singing*, 71(5): 587–94.
Hughes, J. (2010), 'Contradictions from the Enlightenment Roots of Transhumanism', *Journal of Medicine and Philosophy,* 35: 622–40.
Hugill, A. (2008), *The Digital Musician*, London: Routledge.
Husserl, E. (1973), *Experience and Judgement: Investigations in a Genealogy of Logic*, London: Routledge.

IFPI (2010), *IFPI Digital Music Report: Music How, When, Where You Want It*, IFPI.
IFPI (2011), *IFPI Digital Music Report: Music at the Touch of a Button*, IFPI.
IFPI (2015), *Recording Industry in Numbers: The Recorded Music Market in 2014*, IFPI.
Ilan, J. (2012), '"The Industry's the New Road": Crime, Commodification and Street Cultural Tropes in Uk Urban Music', *Crime, Media and Culture*, 8(1): 39–55.
Ingham, T. (2015), 'Major Labels Keep 73% of Spotify Premium Payments', *Music Business Worldwide*, 3 February 2015. http://www.musicbusinessworldwide.com/artists-get-7-of-streaming-cash-labels-take-46/ (accessed 12 January 2016).
Inglis, S. (2003), 'Kieran Hebden of Four Tet is a Producer Who Puts the Intelligence into Intelligent Dance Music', *Sound on Sound*, July 2003. http://www.soundonsound.com/sos/jul03/articles/fourtet.asp (accessed 22 January 2013).
Jacobs, G., and P. Georghiades (1991), *Music and New Technology: The MIDI Connection*, Ammanford: Sigma Press.
Jacobson, W. P. (2011–12), 'Robot's Record: Protecting the Value of Intellectual Property in Music When Automation Drives the Marginal Costs of Music Production to Zero', *Loyola of Los Angeles Entertainment Law Review*, 32(1): 31–46.
Jacobs, G., and P. Georghiades (1991), *Music and New Technology: The Midi Connection*, Wilmslow: Sigma.
Jones, M. L. (2012), *The Music Industries: From Conception to Consumption*, Basingstoke: Palgrave MacMillan.
Jones, S. (1992), *Rock Formation: Music, Technology and Mass Communication*, London: Sage.
Jopson, N. (2015), 'The Economics of Dance', *Resolution*, June 2015, 56–7.
Julien, O. (1999), 'The Diverting of Musical Technology by Rock Musicians: The Example of Double-Tracking', *Popular Music*, 18(3): 357–65.
Katz, M. (2004), *Capturing Sound: How Technology Has Changed Music*, London: University of California Press.
Keeling, R. (2010), 'Shackleton: Man on a String', *Resident Advisor*, Friday, 3 December 2010. http://www.residentadvisor.net/feature.aspx?1263 (accessed 1 December 2011).
Keith, S. (2014), 'Perfect Star, Perfect Style: Lip Synching in J-Pop', in M. Agelucci and C. Caines (eds), *MediaObject 2: Voice/Presence/Absence*, 451–9, Sydney: University of Technology Sydney.
Kelly, C. (2009), *Cracked Media: The Sound of Malfunction*, Cambridge, MA: MIT Press.
Kim-Cohen, S. (2009), *In the Blink of an Ear: Toward a Non-Cochlear Sonic Art*, New York: Continuum.
Kirby, P. (2016), *The Evolution and Decline of the Traditional Recording Studio*, Liverpool: Unpublished PhD Thesis, Liverpool University.
Klein, B., L. M. Meier and D. Powers (2016), 'Selling Out: Musicians, Autonomy, and Compromise in the Digital Age', *Popular Music and Society*, 1–7:1.
Krueger, J. (2010), 'Doing Things with Music', *Phenomenology and the Cognitive Sciences*, 10(1): 1–22.
Kruse, H. (2003), *Site and Sound: Understanding Independent Music Scenes*, New York: Peter Lang Publishing.

Kruse, N. (2012), 'Locating "The Road to Lisdoonvarna" via Autoethnography: Pathways, Barriers and Detours in Self-Directed Online Music Learning', *Journal of Music, Technology and Education*, 5(3): 293–308.

Labelle, B. (2010), *Acoustic Territories: Sound Culture and Everyday Life*, London: Continuum.

Lacasse, S. (2000), *Listen to My Voice: The Evocative Power of Voice in Recorded Rock Music and Other Forms of Vocal Expression*, Liverpool: PhD Thesis, Liverpool University.

Lampel, J., T. Lant and J. Shamsie (2000), 'Balancing Act: Learning from Organizing Practices in Cultural Industries', *Organization Science,* 11(3): 263–9.

Langlois, T. (1992), 'Can You Feel It? DJs and House Music Culture in the UK', *Popular Music*, 11: 229–38.

Latour, B. (1991), 'Technology is Society Made Durable', in J. Law (ed.), *A Sociology of Monsters: Essays on Power, Technology and Domination*, 103–31, London: Routledge.

Latour, B. (1999), *Pandora's Hope. Essays on the Reality of Science Studies*, Cambridge, MA: Harvard University Press.

Latour, B. (2005), *Reassembling the Social: An Introduction to Actor Network Theory,* New York: Oxford University Press.

Ledsham, E. (2015), 'Human After All: Auto-tune Technology and Human Creativity', *Drowned in Sound,* 27 September 2015. http://drownedinsound.com/in_depth/4149415-human-after-all--auto-tune-technology-and-human-creativity (accessed 31 October 2015).

Leenders, M., M. A. Farrell, K. Zwaan and T. Der Bogt (2015), 'How are Young Music Artists Configuring their Media and Sales Platforms in the Digital Age?', *Journal of Marketing Management*, 31(17–18): 1799–817.

Leider, C. (2004), *Digital Audio Workstation: Mixing, Recording and Mastering on Your Mac or PC*, New York: McGraw-Hill.

Leonard, M. (2007), *Gender in the Music Industry: Rock, Discourse and Girl Power*, Aldershot: Ashgate.

Leonard, M. (2010), 'The Creative Process: Liverpool Songwriters on Songwriting', in M. Leonard and R. Strachan (eds), *The Beat Goes On: Liverpool, Popular Music and the Changing City*, 161–80, Liverpool: Liverpool University Press.

Levine M. (2010), 'RedOne: International Hitman', *Electronic Musician*, 9 April 2010. http://emusician.com/interviews/feature/redone_international_hitman/index4.html (accessed 9 September 2011).

Leyshon, A. (2014), *Reformatted: Code, Networks, and the Transformation of the Music Industry*, Oxford, UK: Oxford University Press.

Lobato, R. (2011), 'Constructing the Pirate Audience: On Popular Copyright Critique, Free Culture and Cyber-Libertarianism', *Media International Australia* 139: 11; 113–23.

Lotman, Y. M. (1990), *Universe of the Mind: A Semiotic Theory of Culture,* trans. A. Shukman, London and New York: I. B. Tauris & Co Ltd.

McCarthy, J., and P. Wright (2006), *Technology as Experience*, Cambridge, MA: MIT Press.

McIntyre, P. (2008), 'Creativity and Cultural Production: A Study of Contemporary Western Popular Music Songwriting', *Creativity Research Journal*, 20(1): 40–52.

McIntyre, P., and Paton, B. (2008), 'The Mastering Process and The Systems Model of Creativity', *Perfect Beat,* 8(4): 64–82.

McLean, C. (2014), 'Lady Gaga on Tony Bennett: "I've been controlled for years. He liberates me"', *The Telegraph*, 8 September 2014. http://www.telegraph.co.uk/culture/music/rockandpopfeatures/11075519/Lady-Gaga-on-Tony-Bennett-Ive-been-controlled-for-years.-He-liberates-me.html (accessed 5 January 2016).

McLean, R., P. G. Oliver and D. W. Wainwright (2010), 'The Myths of Empowerment through Information Communication Technologies', *Management Decision*, 48(9): 1365–77.

McLeod, K. (2005), 'Confessions of an Intellectual (Property): Danger Mouse, Mickey Mouse, Sonny Bono, and My Long and Winding Path as a Copyright Activist-Academic', *Popular Music and Society*, 28(1): 79–93.

McLeod, K. (October 2005a), 'MP3s Are Killing Home Taping: The Rise of Internet Distribution and Its Challenge to the Major Label Music Monopoly', *Popular Music and Society,* 28(4): 521–31.

McKinnon, M. (2007), 'South London Calling: This is the Year of Dubstep. Which Means What, Exactly?', *cbc.ca,* 30 January 2007. https://web.archive.org/web/20111208085641/http://www.cbc.ca/arts/music/dub_style.html (accessed 12 December 2015).

McNeil, L. (2012), 'There is No I in Network: Social Networking Sites and Posthuman (Auto)Biography', *Biography*, 35(1): 65–82.

Mace, S. (1985), 'Electronic Orchestras in Your Living Room: MIDI Could Make 1985 the Biggest Year for Computer Musicians', *InfoWorld*, 19 March: 29–33.

Mali, V. V. (2008), 'An Alternative Operating Model for the Record Industry Based on the Development and Application of Non-Traditional Financial Models', *UCLA Entertainment Law Review*, 15(1): 127–37.

Marrington, M. (2011), 'Experiencing Musical Composition in the Daw: The Software Interface as Mediator of the Musical Idea', *Journal on the Art of Record Production*, 05. http://arpjournal.com/experiencing-musical-composition-in-the-daw-the-software-interface-as-mediator-of-the-musical-idea-2/.

Marstal, H. (2002), 'Noise, "noisality", Nostalgia', in T. Millroth (ed.), *Look At the Music/SeeSound*, Copenhagen: Ystads konstmuseum, Neon Gallery, Museet for Samtidskunst.

Martin, L. (2015), 'The VICE Oral History of Dubstep', *Vice.com*, 23 June 2015. http://www.vice.com/read/an-oral-history-of-dubstep-vice-lauren-martin-610/page/0 (accessed 1 December 2015).

Massumi, B. (2002), *Parables for the Virtual: Movement, Affect, Sensation*, London: Duke University Press.

Meelberg, V. (2009), 'Sonic Strokes and Musical Gestures: The Difference between Musical Affect and Musical Emotion', *Proceedings of the 7th Annual Triennial Conference of the European Society for the Cognitive Sciences of Music,* 324–7.

Merlino, M., and C. McCrellis-Mitchell (2015), 'David Guetta: Spinning a New Era in DJ Performance', *the Hub,* 31 March 2015. http://thehub.musiciansfriend.com/artist-interviews/david-guetta-spinning-a-new-era-in-dj-performance (accessed 13 January 2015).

Middleton, R. (1990), *Studying Popular Music,* Milton Keynes: Open University Press.

Millard, D. (2013), 'Is EDM a Real Genre', *Noisey,* 6 May 2013. http://noisey.vice.com/en_ca/blog/is-edm-a-real-genre (accessed February 2016).

Milner, G. (2009), *Perfecting Sound Forever: The Story of Recorded Music*, London: Granta.

Mitchell, G. (August 2013), 'Q and A: Dave Pensado', *Billboard,* 125 (32): 23.

Moore, R. (2007), 'Friends Don't Let Friends Listen to Corporate Rock: Punk as a Field of Cultural Production', *Journal of Contemporary Ethnography,* 46(4): 438–73.

Moorefield, V. (2005), *The Producer as Composer: Shaping the Sounds of Popular Music*, Cambridge, MA: MIT Press.

Monroe, A. (1999), 'Thinking About Mutation: Genres in 1990s Electronica', in A. Blake (ed.), *Living through Pop*, 146–58, London: Routledge.

Mortensen, A. (2012), 'Electronic Dance Music: How Bedroom Beat Boys Remixed the Industry', *CNN.com*, 30 March 2012. http://edition.cnn.com/2012/03/30/showbiz/electronic-dance-music/ (accessed 13 February 2016).

Morton, D. L. (2004), *Sound Recording: The Life Story of a Technology*, Baltimore: Johns Hopkins.

Mulvey, J. M. (1979), 'Strategies in Modeling: A Personal Scheduling Example', *Interfaces*, 9(3): 66–77.

Mulvey, L. (1975), 'Visual Pleasure and Narrative Cinema', *Screen*, 16 (3): 6–18.

Municiple (2006), 'Interview with Deapoh [Bare Files]', *getdarker.com,* 20 April 2006. http://getdarker.com/interviews/interview-deapoh-bare-files/ (accessed 1 December 2015).

Music Radar (2011), 'Skrillex on Ableton Live, Plug-Ins, Production and More', *Music Radar,* 3 November 2011. www.musicradar.com/.../interview-skrillex-on-ableton-live-plug-ins- production-and-more-510973 (accessed 10 February 2016).

Music Trades (November 2014), 'Will EDM's Huge Global Fan Base Lead To Industry Growth?', *Music Trades*, 82–9.

NAMM (2007), *NAMM Global Report: A Statistical Review of the Music Products Industry*, Carlsbad, CA: NAMM.

Nayar, P. (2013), *Posthumanism*, Boston: Polity.

Negus, K. (1992), *Producing Pop: Culture and Conflict in the Popular Music Industry*, London: Edward Arnold.

Negus, K. (1997), *Music Genres and Corporate Cultures,* London: Routledge.

Negus, K. (2002), 'The Work of Cultural Intermediaries and the Enduring Distance between Production and Consumption', *Cultural Studies*, 16 (4): 501–15.

Negus, K., and M. Pickering (2004), *Creativity, Communication and Cultural Value*, London: Sage.

Norman, D. A. (1988), *The Psychology of Everyday Things*, New York: Basic Books.

Norman, D. A. (1999), 'Affordances, Conventions and Design', *Interactions,* 6(3): 38–43.

Nussbaum, C. (2007), *The Musical Representation: Meaning, Ontology, and Emotion,* Cambridge, MA: MIT Press.

O'Malley Greenburg, Z., and N. Messitte (2015), 'Revenge of the Record Labels: How the Majors Renewed their Grip on Music', *Forbes,* 15 April 2015. http://www.forbes.com/sites/zackomalleygreenburg/2015/04/15/revenge-of-the-record-labels-how-the-majors-renewed-their-grip-on-music/2/ (accessed 23 April 2015).

O'Neil, T. (2006), 'Ryoji Ikeda: Dataplex Review', *Popmatters*. http://www.popmatters.com/pm/review/ryoji_ikeda_dataplex/.

OED (2006), *Oxford Dictionary of English, Second Edition Revised*, Oxford: Oxford University Press.

OED (2015a), 'Digitization, n.', *OED Online*, Oxford University Press, December 2015. http://www.oed.com.liverpool.idm.oclc.org/view/Entry/240886?redirectedFrom=digitization (accessed16 February 2016).

OED (2015b), 'Digitalization, n.2.', *OED Online*, Oxford University Press, December 2015. http://www.oed.com.liverpool.idm.oclc.org/view/Entry/242061?rskey=A6D0EP&result=2 (accessed16 February 2016).

Oliveira, A. L. G., and Oliveira, L. F. (2002), *Toward an Ecological Conception of Timbre*, Unpublished Conference Paper.

Parikka, J. (2013), 'Introduction', in W. Ernst (ed.), *Digital Memory and the Archive,* 1–22, Minneapolis: University of Minnesota Press.

Palfrey, J., and U. Gasser (2008), *Born Digital: Understanding the First Generation of Digital Natives*, New York: Basic Books.

Pea, R. (1993), 'Practices of Distributed Intelligence and Designs for Education', in G. Salomon (ed.), *Distributed Cognition: Psychological and Educational Considerations*, 47–87, Cambridge: Cambridge University Press.

Perry, M. (2010), *How to Be a Record Producer in the Digital Era,* New York: Billboard.

Peterson, R. (1997), *Creating Country Music: Fabricating Authenticity*, Chicago: University of Chicago Press.

Pettiman, D. (2012), 'Pavlov's Podcast: The Acousmatic Voice in the Age of MP3s', *Differences: Journal of Feminist Cultural Studies*, 22(4–5): 140–67.

Pickering, J. (2007), 'Affordances are Signs', *TripleC,* 5(2): 64–74.

Piekut, B. (2014), 'Actor-Networks in Music History: Clarifcations and Critiques', *Twentieth-Century Music*, 11: 191–215.

Piore, M. J., and C. F. Sabel (1984), *The Second Industrial Divide*, New York: Basic Books.

Platz, R. (1995), 'More than Just Notes: Psychoacoustics and Composition', *Leonardo Music Journal*, 5: 23–8.

Pratt, A. C. (1997), 'The Cultural Industries Production System: A Case Study of Employmentchange in Britain, 1984–91', *Environment and Planning*, 29(11): 1953–74.

Press, L. (1993), 'Before the Altair: The History of Personal Computing', *Communications of the ACM*, 36(9): 27–33.

Prior, N. (2008a), 'OK COMPUTER: Mobility, Software and the Laptop Musician', *Information, Communication and Society*, 11(7): 912–32.

Prior, N. (2008b), 'Putting a Glitch in the Field: Bourdieu, Actor Network Theory and Contemporary Music', *Cultural Sociology*, 2(3): 301–19.

Prior, N. (2009), 'Software Sequencers and Cyborg Singers: Popular Music in the Digital Hypermodern', *New Formations*, 66: 81–99.

Prior, N. (2010), 'The Rise of the New Amateurs: Popular Music, Digital Technology and the Fate of Cultural Production', in John R. Hall, L. Grindstaff and M.-C. Lo (eds), *Handbook of Cultural Sociology*, 398–407, London: Routledge.

Rietveld, H. (1998), *This is Our House: House Music, Cultural Spaces and Technologies,* Aldershot: Ashgate.

Rehn, A. (2004), 'The Politics of Contraband: The Honor Economies of the Warez Scene', *Journal of Socio-Economics,* 33: 359–74.

Reyes, I. (2010), 'To Know Beyond Listening: Monitoring Digital Music', *The Senses and Society,* 6 (2): 322–38.

Reynolds, N. (2008), 'Seeking Affordances: Searching for New Definitions and New Understandings of Children's Relationships with Technologies in Musical Compositions', *Australian Educational Computing,* 23(1): 15–18.

Reynolds, S. (1998), *Energy Flash: A Journey Through Rave Music and Dance Culture*, London: Picador.

Reynolds, S. (2010), *Retromania: Pop Culture's Addiction to its Own Past*, London: Faber and Faber.

Ridge, P. (2010), 'Interview – Trentemøller: a Quantum Leap', *Summer Festival Guide 2010.* http://summerfestivalguide.com.au/SPA_summer_fest/archives/2798 (accessed 22 February 2011).

Rinehart (2008), 'Interrogating Martyn', *Resident Advisor*, 18 June 2008. http://www.residentadvisor.net/feature.aspx?925 downloaded (accessed 5 June 2013).

Richardson, I. (2010), 'Faces, Interfaces, Screens: Relational Ontologies of Framing, Attention', *Transformations,* 18. http://www.transformationsjournal.org/journal/issue_18/article_05.shtml (accessed 1 May 2011).

Rogers, J. (2009), 'Top of their Voices: From accents to Auto-Tune, Singers Fought to Stand Out From the Pack', *New Statesman,* 14 December 2009, 47–8.

Rogers, J. (2013), *The Death and Life of the Music Industry in the Digital Age*, New York: Bloomsbury.

Rollefson, J. G. (2008), 'The "Robot Voodoo Power" Thesis: Afrofuturism and Anti-Anti-Essentialism from Sun Ra to Kool Keith', *Black Music Research Journal*, 28(1): 83–109.

Rudolf, T., and J. Frankel (2009), *Youtube in Music Education*, New York: Hal Leonard.

Russ, M. (2004), *Sound Synthesis and Sampling: Second Edition*, Burlington, MA: Focal Press.

Rutsky, R. L. (1999), *High Techne: Art and Technology From the Machine Aesthetic to the Posthuman*, Minneapolis, MN: University of Minnesota Press.

Ryan, J., and M. Hughes (2006), 'Breaking the Decision Chain: The Fate of Creativity in the Age of Self-Production', in M. D. Ayers (ed.), *Cybersounds: Essays on Virtual Music Culture*, 239–53, New York: Peter Lang.

Salavuo, M. (2006), 'Open and Informal Online Communities as Forums of Collaborative Musical Activities and Learning', *British Journal of Music Education*, 23(3): 253–71.

Salvato, N. (2009), 'Out of Hand: Youtube Amateurs and Professionals', *TDR: The Drama Review*, 53(3): 67–83.

Sandywell, B., and D. Beer (2005), 'Stylistic Morphing: Notes on the Digitisation of Contemporary Music Culture', *Convergence,* 11(4): 106–21.

Sangild, T. (2004), 'Glitch—The Beauty of Malfunction', in C. Washburne and M. Derno (eds), *Bad Music: The Music We Love to Hate,* 257–74, London: Routledge.

Savage, S. (2006), 'Ignorance is Bliss: Capturing the Unintentional Performance', *Popular Music and Society*, 18(3): 332–40.

Schaeffer, P. (1966), *Traite´ des objets musicaux*, Paris: Éditions du Seuil.

Schafer, R. M. (1977), *The Soundscape: Our Sonic Environment and the Tuning of the World*, Rochester: Destiny.

Schellenberg, E. G., and S. E. Trehub (1996), 'Natural Musical Intervals', *Psychological Science*, 7(5): 272–7.

Schmidt Horning, S. (2004), 'Engineering the Performance: Recording Engineers, Tacit Knowledge and the Art of Controlling Sound', *Social Studies of Science,* 34(5): 703–31.

Scott, M. (2012), 'Cultural Entrepreneurs, Cultural Entrepreneurship: Music Producers Mobilising and Converting Bourdieu's Alternative Capitals', *Poetics,* 40: 237–55.

Senior, M. (2011), *Mixing Secrets for the Small Studio*, Amsterdam: Elsevier/Focal Press.

Serazio, M. (2008), 'The Apolitical Irony of Generation Mash-up: A Cultural Case Study in Popular Music', *Popular Music and Society,* 31(1): 79–4.

Sexton, J. (2012), 'Weird Britain in Exile: Ghost Box, Hauntology, and Alternative Heritage', *Popular Music and Society,* 35(4): 561–84.

Sherburne, P. (2002), '12k: Between Two Points', *Organised Sound,* 7(1): 171–6.

Shiga, J. (2007), 'Copy-and-Persist: The Logic of Mash-up Culture', *Critical Studies in Media Communication,* 24(2): 93–114.

Simonton, D. K. (1999a), *Origins of Genius: Darwinian Perspectives on Creativity*, New York: Oxford University Press.

Sisario, B. (2014), 'Indie Music's Digital Drag: Small Music Labels See YouTube Battle as Part of War for Revenue', *New York Times,* 24 June 2014. http://www.nytimes.com/2014/06/25/business/media/small-music-labels-see-youtube-battle-as-part-of-war-for-revenue.html?_r=0 (accessed 20 February 2016).

Small, C. (1998), *Musicking: The Meanings of Performing and Listening*, Hannover, NH: Wesleyan University Press.

Sonic Academy (2012), '1'5 Questions With Shadow Dancer', *Sonic Academy*, 17 April 2012. http://www.sonicacademy.com/Producer+Interviews/articles/15-Questions-With-Shadow-Dancer–An-Interview-With-Shadow-Dancer.cid5136 (accessed 12 January 2016).

Stahl, M., and L. M. Meier. (2012), 'The Firm Foundation of Organizational Flexibility: The 360 Contract in the Digitalizing Music Industry', *Canadian Journal of Communication*, 37(3): 441–58.

Standley, J., and C. Madsen (1990), 'Comparison of Infant Preferences and Responses to Auditory Stimuli: Music, Mother, and Other Female Voice', *Journal of Music Therapy*, 27: 54–97.

Sterne, J. (2006), *The Audible Past: Cultural Origins of Sound Reproduction*, London: Duke University Press.

Sterne, J. (2006a), 'The mp3 as Cultural Artifact', *New Media and Society,* 8(5): 825–42.

Stobart, H. (2010), 'Rampant Reproduction and Digital Democracy: Shifting Landscapes of Music Production and "Piracy" in Bolivia', *Ethnomusicology Forum*, 19 (1): 27–56.

Strachan, R. (2003), *Do-It-Yourself: Industry, Ideology, Aesthetics and Micro Independent Record Labels in the UK*, Liverpool: Unpublished PhD Thesis, Liverpool University.

Strachan, R. (2007), 'Micro-Independent Record Labels in the UK. Discourse, DIY Cultural Production and the Music Industry', *European Journal of Cultural Studies*, 10(2): 245–65.

Strachan, R. (2010), 'Uncanny Space: Theory, Experience and Affect in Contemporary Electronic Music', *Trans: Transcultural Music Review*, Vol. 13,

http://www.sibetrans.com/trans/articulo/14/uncanny-space-theory-experience-and-affect-in-contemporary-electronic-music
Strachan, R. (2012), 'Affordances, Stations Audionumeriques et Creativity Musicale', *Reseaux: Communication – Technologie - Société*, 172: 120–43.
Strachan, R. (2013), 'The Spectacular Suburb: Creativity and affordance in Contemporary Electronic Music and Sound Art', *SoundEffects – An Interdisciplinary Journal of Sound and Sound Experience,* 2(3): 7–25.
Stratton, J. (1983), 'Capitalism and Romantic Ideology in the Record Business', *Popular Music,* 3: 143–56.
Straw, W. (1991), 'Systems of Articulation, Logics of Change: Communities and Scenes in Popular Music', *Cultural Studies,* 5(3): 368–88.
Swash, R. (2011), 'Music Industry Dances to Technology's Tune', *Guardian*, 17 March 2011. http://www.theguardian.com/culture/2011/mar/17/sxsw-music-technology-soundcloud (accessed 12 January 2015).
Tagg, P. (1991), 'Towards a Sign Typology of Music', in R. Dalmonte and M. Baroni (eds), *Secondo convegno europeo di analisi musicale,* 369–78, Trento: Universita Degli Studi Trento.
Tagg, P. (1994), 'From Refrain to Rave: The Decline of Figure and the Rise of Ground', *Popular Music*, 13(2): 209–22.
Tarasti, E. (2002), *Signs of Music: A Guide to Musical Semiotics*, Berlin: Mouton de Gruyter.
Tavana, A. (2015), 'Democracy of Sound: Is Garageband Good For Music?', *Pitchfork,* 30 September 2015. http://pitchfork.com/features/articles/9728-democracy-of-sound-is-garageband-good-for-music/ (accessed1 February 2016).
Taylor, T. (2001), Strange Sounds: Music, Technology and Culture, London: Routledge.
Théberge, P. (1997), *Any Sound You Can Imagine: Making Music/Consuming Technology,* Hanover, NH: Wesleyan University Press.
Théberge, P. (2004), 'The Network Studio: Historical and Technological Paths to a New Ideal in Music Making', *Social Studies of Science*, 34(5): 759–81.
Théberge, P. (2012), '"The End of the World as We Know It" The Changing Role of the Studio in the Age of the Internet', In S. Z. Thomas and S. Frith (eds), *The Art of Record Production*, 77–90, Farnham: Ashgate.
Théberge, P. (2015), 'Digitalization', in J. Shepherd and K. Devine (eds), *The Routledge Reader on the Sociology of Music*, 329–40, London: Routlege.
Thompson, J. (2011), 'Djent, the Metal Geek's Microgenre', *The Guardian,* 3 March 2011. http://www.theguardian.com/music/2011/mar/03/djent-metal-geeks (accessed 19 March 2013).
Thornton, S. (1995), *Club Cultures: Music, Media and Subcultural Capital*, Cambridge: Polity Press.
Thrift, N., and S. French (2002), 'The Automatic Production of Space', *Transactions of the Institute of British Geographers,* 27(3): 309–35.
Tingen, P. (2010), 'Secrets of the Mix Engineers: Greg Kurstin', *Sound on Sound,* March 2010. http://www.soundonsound.com/sos/mar10/articles/it_0310.htm (accessed 10 February 2016).
Tingen, P. (September 2014), 'Inside Track, Secrets of the Mix Engineers: Steve Fitzmaurice', *Sound on Sound,* 38–51.
Tingen, P. (December 2015), 'Inside Track, Secrets of the Mix Engineers: Carlo "Illangelo" Montagnese', *Sound on Sound*, 39–52.

Toffler, A. (1970), *Future Shock*, New York: Random House.
Tost, M. (2015), 'Avicii, Martin Garrix, and Others Have Used Pirated Software'. http://www.youredm.com/2015/02/24/avicii-martin-garrix-others-use-pirated-software/ (accessed 5 March 2015).
Toynbee, J. (2000), *Making Popular Music*, London: Arnold.
Toynbee, J. (2002), 'Mainstreaming: From Hegemonic Centre to Global Networks', in D. Hesmondhalgh and K. Negus (eds), *Popular Music Studies*, 149–63, London: Hodder Arnold.
Trainor, L. J., and B. M. Heinmiller (1998), 'The Development of Evaluative Responses to Music: Infants Prefer to Listen to Consonance Over Dissonance, *Infant Behavior and Development*, 21(1): 77–88.
Tschmuck, P. (2012), *Creativity and Innovation in the Music Industry*, Heidelberg: Springer.
Tucker, S. (2003), *The Secrets of Songwriting: Leading Songwriters Reveal How to Find Inspiration and Success*, New York: Skyhorse.
Turner, F. (2006), *From Counterculture to Cyberculture: Stewart Brand, the Whole Earth Network, and the Rise of Digital Utopianism*, Chicago: University of Chicago Press
Turner, P. (2008), 'Towards an Account of Intuitiveness', *Behaviour and Information Technology*, 27(6): 475–82.
Vale, V., and A. Juno (1993), *Incredibly Strange Music Vol I*, San Francisco: Re:Search.
Vallee, M. (2013), 'The Media Contingencies of Generation Mashup: A Zˇ izˇekian Critique', *Popular Music and Society*, 36(1), 76–97.
Van Arman, D. (2015), 'Testimony of Darius Van Arman Co-Founder of Secretly Group Bloomington, Indiana, Board of Director member of the American Association of Independent Music ("A2IM") before the United States of America House of Representatives House Judiciary Committee. Subcommittee on Courts, Intellectual Property and the Internet.' http://judiciary.house.gov/_cache/files/0f007c39–4b39-4604-8c62-79e58af436a8/final-a2imdariusvanarman0621.pdf (accessed 14 December 2015).
Vanhanen, J. (2003), 'Virtual Sound: Examining Glitch and Production', *Contemporary Music Review*, 22(4): 45–52.
Veal, M. (2007), *Dub: Soundscapes and Shattered Songs in Jamaican Reggae*, Middletown, CT: Wesleyan.
Visell, Y. (2009), 'Tactile Sensory Substitution: Models for Enaction in HCI', *Interacting with Computers*, 21: 38–53.
Von Appen, R. (2015), 'What Makes Ke$ha's Tik Tok Tick', in R. von Appen and A. Doehring (eds), *Song Interpretation in 21st-Century Pop Music*, 29–52, Aldershot: Ashgate.
Waldron, J. (2009), 'Exploring a Virtual Music "community of practice": Informal Music Learning on the Internet', *Journal of Music, Technology and Education*, 2(3): 97–112.
Walser, R. (1993), *Running with the Devil: Power, Gender, and Madness in Heavy Metal Music*, Hannover, NH: Wesleyan University Press.
Warner, T. (2003), *Pop Music Technology and Creativity: Trevor Horn and the Digital Revolution*, Aldershot: Ashgate.
Watson, K. (2014), *IMS Business Report 2014*, Ibiza: International Music Summit.
Wenger, E. (1998), *Communities of Practice: Learning, Meaning, and Doing*, New York: Cambridge University Press.

Whelan, A. (2006), 'Do U Produce?: Subcultural Capital and Amateur Musicianship in Peer-to Peer Networks', in M. D. Ayers (ed.), *Cybersounds: Essays on Virtual Music Culture*, 57–82, New York: Peter Lang.

Weheliye, A. G. (2002), '"Feenin": Posthuman Voices in Contemporary Black Popular Music', *Social Text*, 20(2): 21–47.

Wikström, P. (2009), *The Music Industry*, London: Polity.

Williams, A. (2012), 'Putting It On Display: The Impact Of Visual Information On Control Room Dynamics', *Journal on the Art of Record Production*, 06. http://arpjournal.com/putting-it-on-display-the-impact-of-visual-information-on-control-room-dynamics/ (accessed 05 May 2016).

Windsor, L. (2000), 'Through and Around the Acousmatic: the Interpretation of Electroacoustic Sounds', in S. Emmerson (ed.), *Music, Electronic Media and Culture*, 7–35, London: Ashgate.

Windsor, L. (2004), 'An Ecological Approach to Semiotics', *Journal for the Theory of Social Behaviour,* 34(2).

Wishart, T. (1986), 'Sound Symbols and Landscapes', in S. Emerson (ed.), *The Language of Electroacoustic Music,* 41–60, London: MacMillan.

Wolfe, C. (2009), *What Is Posthumanism?* Minneapolis, MN: University of Minnesota Press.

Wren and Reynolds (2004), 'Minimalism in ubiquitous interface design', *Ubiquitous Computing,* 8: 370–3.

Young, R. (1999, December/2000, January), Undercurrents #12: Worship the glitch, *The Wire: Adventures in Modern Music*, 190(191): 52–6.

Zak, A. J. (2001), *The Poetics of Rock: Cutting Tracks, Making Records*, Berkley: University of California Press.

Zagorski-Thomas, S. (2013), 'Real and Unreal Performances: The Interaction of Recording Technology and Rock Drum Kit Performance', in A. Danielsen (ed.), *Musical Rhythm in the Age of Digital Reproduction*, 195–212, Farnham: Ashgate.

Zagorski-Thomas, S. (2014), *The Musicology of Record Production*, Oxford: Oxford University Press.

Zolberg, V. (1990), *Constructing a Sociology of the Arts*, Cambridge: Cambridge University Press.

Zollo, P. (1997), *Songwriters on Songwriting*, New York: Da Capo Press.

INDEX